Cognitive Decline with Ultra-Processed Food

COGNITIVE DECLINE WITH ULTRA-PROCESSED FOOD

Written by:

Austin Mardon
Muzammil Bin Younus
Ayushma Neku
Maureen Saha
Wan Ling Dai
Michael Tang
Ipsa Gusain
Lydia Clarke Rehman
Uzair Tazeem

Edited by:

Catherine Mardon
Peter Anto Johnson
John Christy Johnson

GM★ PRESS
2022

A Golden Meteorite Press Book.
Printed in Canada.

First Printing: 2022

Typeset and Cover Design by Fariha Khan

ISBN: 978-1-77369-870-0

Golden Meteorite Press
103 11919 82 St NW
Edmonton, AB T5B 2W3
www.goldenmeteoritepress.com

TABLE OF CONTENTS

INTRODUCTION TO COGNITIVE FUNCTIONS AND THE PROCESSED FOOD INDUSTRY   1
Muzammil Bin Younus

HISTORY AND TIMELINE OF STUDYING COGNITIVE DECLINE AND PROCESSED FOOD   10
Ayushma Neku

SCIENCE   21
Kelly Wu

ECONOMICS   29
Wan Ling Dai

POLITICS   37
Michael Tang

IMPLICATIONS   46
Ipsa Gusain

THE ROLE OF HEALTH AND ENVIRONMENTAL ETHICS WITHIN THE PROCESSED FOOD INDUSTRY   56
Lydia Clarke Rehman

PRACTICAL ACTIONS TO ASSISTING COGNITIVE DECLINE AND HEALTHY HABITS   64
Uzair Tazeem

REFERENCES   74

1

INTRODUCTION TO COGNITIVE FUNCTIONS AND THE PROCESSED FOOD INDUSTRY

By Muzammil Younus

INTRODUCTION TO PROCESSED FOODS

The U.S. Department of Agriculture (USDA) defines a processed food as any raw agricultural commodity that has undergone any changes to its natural state (Sadler et al., 2021). This includes the raw commodity that has been subjected to washing, cleaning, milling, cutting, chopping, heating, pasteurizing, blanching, cooking, canning, freezing, drying, dehydrating, mixing, packaging, and other processes (Parrish, 2014). It may also include addition of other ingredients to the food, such as preservatives, flavors, nutrients and other food additives (Parrish, 2014).

Food processing industry plays an important role in providing edible, safe and nutritious food to the world's population (Sadler et al., 2021). The food and beverage processing industry is the second largest manufacturing industry in Canada, producing goods worth

$117.8 billion in 2019. It accounts for 17% of the total manufacturing sales and 2% of the national Gross Domestic Product (GDP), with a labor force of 290,000 (Overview of the food and beverage processing industry - agriculture.canada.ca, n.d.).

However, the topic of food processing is a complicated one due to the many types of processes involved, that may bring both benefits and risks depending on the context (Sadler et al., 2021). For example, heat treatment which is used to increase shelf life may also reduce food-borne illness and improve digestibility. However, thermal processing can also have undesirable consequences such as loss of certain nutrients and formation of toxic compounds. For example, processed meats may contain heterocyclic amines and polycyclic aromatic hydrocarbons which are carcinogenic chemicals, and acrylamide, which can cause nerve damage to humans (Sadler et al., 2021).

There are many processed food classification systems, however the classification criteria used are inconsistent (Sadler et al., 2021). A popular processed food classification system is NOVA classification, which was introduced in 2009 and is recognized by various organizations such as World Health Organization and Food and Agriculture Organization (Processed Foods and Health, n.d.). NOVA categorizations divide foods into four categories; Unprocessed and Minimally Processed Foods, Processed Culinary Ingredients, Processed Food Products, Ultra-processed Products (Sadler et al., 2021).

Unprocessed foods are the edible parts of plants and of or from animals (Food and Agriculture Organization of the United Nations, 2019). These include fruits, vegetables, grains, legumes, meat, poultry, seafood amongst many other naturally occurring edible items (Food and Agriculture Organization of the United Nations, 2019). Fungi, algae and water are also unprocessed foods (Food and Agriculture Organization of the United Nations, 2019).

Minimally processed foods are natural or unprocessed foods altered by processes such as removal of inedible parts, drying, powdering, squeezing, crushing, grinding, chilling, pasteurization, freezing and other processes that do not add salt, sugar, oils or fats

2

or any other food substances to the original food. The purpose of these processes is to preserve natural foods to extend their life and make them suitable for storage, or to make them safe, edible and/or more pleasant to consume, or to make their preparation more diverse (Food and Agriculture Organization of the United Nations, 2019).

Unprocessed and minimally processed foods may be combined together, or may be prepared and cooked as meals in combination with processed culinary ingredients and sometimes with some processed foods (Food and Agriculture Organization of the United Nations, 2019).

The second food category is processed culinary ingredients which include substances obtained from unprocessed or from nature by industrial processes such as pressing, centrifuging, refining, extracting or mining. These include oils, butter, lard, sugar and salt (Food and Agriculture Organization of the United Nations, 2019). These are rarely consumed by themselves, but instead are used to prepare, season and cook unprocessed and minimally processed foods (Food and Agriculture Organization of the United Nations, 2019). They are also energy-dense with the exception of salt, and are at 400-900 kilocalories per 100 grams (Food and Agriculture Organization of the United Nations, 2019). This is 3-6 times more than cooked grains and around 10-20 times more than cooked vegetables (Food and Agriculture Organization of the United Nations, 2019).

The third food category is processed foods which are products made by adding salt, oil, sugar or other processed culinary ingredients to unprocessed and minimally processed foods (Food and Agriculture Organization of the United Nations, 2019). Processed foods preparation includes various preservation processes such as canning and bottling (Food and Agriculture Organization of the United Nations, 2019). These processes are designed to increase the durability of the food and make them more enjoyable by modifying or enhancing their sensory qualities (Food and Agriculture Organization of the United Nations, 2019). Examples of processed foods include canned or bottled vegetables and legumes preserved in brine, fruit preserved in syrup, tinned fish preserved in oil, processed animal food such as ham, bacon and smoked fish, most baked breads, cheeses and salted nuts and seeds (Food and Agriculture Organization of the United Nations, 2019).

Processed foods have two or more ingredients and are recognisable as modified versions of unprocessed foods (Food and Agriculture Organization of the United Nations, 2019). However, they usually retain the basic identity and most constituents of the original food (Food and Agriculture Organization of the United Nations, 2019). Except for canned vegetables, their energy density ranges from moderate (around 150-250 kilocalories per 100 grams for most processed meats) to high (around 300-400 kilocalories per 100 grams for cheeses) (Food and Agriculture Organization of the United Nations, 2019).

The fourth food category is ultra-processed foods which are made by a series of industrial processes, many requiring sophisticated equipment and technology (Food and Agriculture Organization of the United Nations, 2019). These processes include fractioning of whole foods into substances, chemical modification of these substances, assembly of unmodified and modified food substances using industrial technique, use of additives at various stages of the manufacturing process, and sophisticated packaging, usually with plastic and other synthetic materials (Food and Agriculture Organization of the United Nations, 2019).

Ingredients of ultra-processed foods include those which have no or rare culinary use such as fructose corn syrup, protein isolates, flavor enhancers, colors, emulsifiers, sweeteners, thickeners and anti-foaming, bulking, carbonating, foaming, gelling and glazing agents (Food and Agriculture Organization of the United Nations, 2019). The processes and ingredients used to manufacture the ultra-processed foods are designed to create highly profitable products which are convenient (ready to consume) and hyper-palatable to displace freshly prepared meals from all other food categories (Food and Agriculture Organization of the United Nations, 2019). Examples of ultra-processed foods include carbonated soft drinks, chocolate, candies, ice cream, cereals, energy bars, energy drinks, milk drinks, fruit drinks, ready to heat products such as pasta and pizza, nuggets, sausages and burgers, and instant noodles. Infant formulas and other baby products are also considered ultra-processed foods (Food and Agriculture Organization of the United Nations, 2019).

REPUTATION OF PROCESSED FOODS

Recent research shows that 43% of consumers have a negative view of processed foods, whereas only 18% have a positive view (Fox, 2012). Processed foods are frequently criticised as unhealthy and environmentally unsustainable, while also being responsible for 'McDonaldization' of food, which refers to erosion of distinctiveness of local food-ways as part of cultural homogenization (Jackson et al., 2018). For example, a study published by British Medical Journal reported that none of the 100 ready to eat meals it tested conformed to the World Health Organization's dietary standards, while another report from UK Department for Environment, Food and Rural Affairs reported that ready to eat food had high greenhouse gas emissions and heavy transport costs, consuming large amount of energy, land and water (Jackson et al., 2018).

A study conducted in New Zealand reported that ultra-processed foods have the worst nutrient profile, yet they are the most available packaged food in a sample of supermarkets (Luiten et al., 2015). The analyses performed by the study showed a positive association between the level of industrial processing and the nutrient profiling score, indicating that ultra-processed foods had a worst nutrient profile than culinary processed foods, which in turn had a worse nutrient profile than minimally processed foods (Luiten et al., 2015). While these foods may contain some of unprocessed and minimally processed food ingredients, they are mostly based on low nutrient density, little dietary fibre and excess simple carbohydrates, saturated fats, sodium and trans fatty acids (Monteiro, 2009). Despite this, a large majority, 83%, of the packaged foods in the sample of supermarkets were of the ultra–processed classification. The high profitability of high availability of ultra-processed foods can be credited for the high availability of these foods (Luiten et al., 2015).

Experimental studies have also indicated that ultra-processed foods induce high glycaemic responses and have satiety potential, and also create a gut environment that selects microbes that promote diverse forms of inflammatory disease (Monteiro et al., 2019). Studies have also shown correlation between increased intake of ultra-processed diet and increased obesity, hypertension, coronary and cerebrovascular diseases, and many other diseases (Monteiro et al., 2019). According to another study, a 10% increase in the proportion of ultra-processed food was associated with a greater than 10% increase in risks of overall and breast cancer (Fiolet et al., 2018). Furthermore, increase in ultra-processed food can also be linked with the development of cognitive decline.

NEURODEGENERATIVE DISEASES AND COGNITIVE DECLINE

Neurodegenerative diseases are characterised by the loss of neurons (Martínez Leo & Segura Campos, 2020). Progressive neurodegeneration leads to the presence of symptoms that include balance impairment, movement (ataxia), speech, breathing, heart functions, and cognitive decline (dementia) (Martínez Leo & Segura Campos, 2020). "According to the National Institute of Neurologic Disorders and Stroke, the most prevalent Neurodegenerative Diseases are Alzheimer's disease, Parkinson's disease, amyotrophic lateral sclerosis, Friedreich's Ataxia, Huntington disease, and spinal muscular atrophy" (Martínez Leo & Segura Campos, 2020).

Old age is the greatest risk factor for most neurodegenerative diseases (Wahl et al., 2016). The ageing of our societies due to increase in average age is leading to a dramatic increase in the prevalence of chronic conditions, which threatens our healthcare systems. In particular, dementia has become a global concern, placing a significant financial and social burden on patients, carers and the health care system (Chen et al., 2019). The significance of this can be seen from the numbers associated with Alzheimer's disease, which is the most common form of dementia. It currently affects 5 million people in the United States, which is projected to rise to approximately 14 million by 2050 (Martin et al., 2015).

There is currently no effective treatment to modify the course of dementia and therefore prevention is an urgent priority, to reduce incidence as well as to slow down progression (van de Rest et al., 2015). Therefore, efforts have been made to identify important risk factors and, in particular, factors that can be modified (van de Rest et al., 2015).

Psychological, lifestyle, education, social networking and cardio-vascular risk factors have all been linked to cognitive health among older adults (Chen et al., 2019). It is estimated that approximately 35% of dementia cases may be due to modifiable environmental factors (Weinstein et al., 2022). Studies have indicated that healthy lifestyles, which includes healthy diet and physical activity, can improve cognitive health and even modestly delay dementia (Weinstein et al., 2022). The role of a healthy diet in maintaining brain health and mitigating cognitive decline has been studied, and several evidence suggests that certain dietary patterns have shown to slow down cognitive decline. These dietary patterns are characterised by high intake of fruits and vegetables, and are low in unhealthy ingredients such as excess sodium, saturated fats and added sugar (Weinstein et al., 2022). These healthier diets are also low in ultra-processed foods (Weinstein et al., 2022).

Study conducted by School of Public Health, University of Haifa in Israel showed that western-style dietary pattern, which is characterised by high Ultra-processed foods is associated with early markers of Alzheimer's disease, whereas, higher quality diet such as Mediterranean diet which have less ultra-processed content, have shown to promote promote brain health (Weinstein et al., 2022). The findings of this study and a secondary analysis from a large cohort study imply that high consumptions of processed meat and oils/spreads such as margarine may be associated with cognitive decline and are a strong predictor of Alzheimer's disease (Weinstein et al., 2022).

ULTRA-PROCESSED FOODS AND COGNITIVE DECLINE

As mentioned above, diet is an environmental factor that can impact cognitive health. Diet is the main factor that influences the diversity and functionality of the gut microbiota (Martínez Leo & Segura Campos, 2020). Microbiota has an important function in the generation of compounds that participate in the energy and intermediary metabolism of organs including the brain, with which it keeps close communication (Martínez Leo & Segura Campos,

2020). There is bidirectional communication between the gut and the brain in which the microbiota, the generic nervous system, the autonomic nervous system, the neuroendocrine system, the neuroimmune system and the Central Nervous System participate (Martínez Leo & Segura Campos, 2020).

Alternations of the diversity and composition of the microbiota is known as dysbiosis (Martínez Leo & Segura Campos, 2020). Changes in dietary pattern with more ultra-processed foods is characterised by an increase in intake of simple carbohydrates and saturated fatty acids, not only leads to changes in gut microbiota composition (dysbiosis) but also in the proinflammatory state related to the development of neurodegenerative diseases (Martínez Leo & Segura Campos, 2020). In other words, high fat and simple carbohydrate diets, such as consumption of chronic ultra-processed foods, is associated with a reduction in cognitive function (Martínez Leo & Segura Campos, 2020).

Several studies in animals have also shown the effects of high-fat diets on ageing in microglial function, cognitive decline and neuroinflammation, as well as contribution towards neurodegenerative diseases such as Alzheimer's disease (Martínez Leo & Segura Campos, 2020). Another study by Wu and Sun with 34,168 participants concluded that the mediterranean diet is inversely related to the risk for incidence of cognitive disorder (Martínez Leo & Segura Campos, 2020), which is in line with results from study conducted by University of Haifa as discussed above. Study by Morris et al. also supported the link between fibre consumption and positive impact on cognitive function; the study concluded that a diet consisting of approximately 1 serving of green leafy vegetables per day and foods rich in phylloquinone, lutein, nitrate, folate, α-tocopherol, and kaempferol were associated with slower cognitive decline (Martínez Leo & Segura Campos, 2020).

Although further studies are needed to confirm whether diet is the main trigger for dysbiosis, studies have made it clear that dietary components such as fibre and antioxidants contribute to gut microbiota homeostasis, which reduces neurodegenerative effects in the nervous system (Martínez Leo & Segura Campos, 2020).

Study in (Martínez Leo & Segura Campos, 2020) concluded that the food industry is increasingly providing access to food with lower nutritional quality and that is affordable, which due to current global food crises are mostly consumed by the population. Furthermore, current lifestyles are driving the population to an increasingly industrialised and less natural diet, which contributes to higher prevalence of chronic metabolic diseases, which in turn are related to neuroinflammatory states, neurodegeneration and cognitive deterioration.

With the result of studies discussed above, it can be said that gut microbiota dysregulation due to increased consumption of ultra-processed foods is a possible reason for neurodegenerative diseases and cognitive decline. Although further studies are required to confirm this relationship, current studies are encouraging and have provided a path forward for further research to confirm the role of healthier diet in modifying the course of dementia or even preventing it.

2

HISTORY AND TIMELINE OF STUDYING COGNITIVE DECLINE AND PROCESSED FOOD

By Ayushma Neku

EVOLUTION OF PROCESSED FOOD AND NUTRITIONAL STUDIES

The use of fire as a basic food processing mechanism, dating back to the Early and Middle Pleistocene approximately 1.8 million years ago in South Africa, provided a basis for the evolution of human nutrition (Baker et al., 2020; Huebbe & Rimbach, 2020). In the Neolithic Era approximately 12 thousand years ago, foraged and cultivated foods were converted into forms more palatable, safe, and long-lasting through cutting, grinding, drying, salting, fermenting, and smoking (Baker et al., 2020). This enabled hunter-gatherers to thrive in various environments, transitioning into an era of agriculture and settlement. Another major transition driving the evolution of food processing was the Industrial Revolution (Huebbe & Rimbach, 2020). The invention of electricity, mass production of steel, and machine manufacturing during this time led to various advancements in food processing. For instance, steam and rolling mills allowed for refined flour production. Canned and refrigerated meat, butter, and vegetables were mass-produced and traded globally alongside non-perishables such as sugar, cocoa, tea, and coffee (Baker et al., 2020). However, from the late 19th and early 20th centuries, the refinement of grain milling led to populations suffering from nutrient deficiencies (Huebbe & Rimbach, 2020). While food and nutrition had been previously studied, this gave rise to the modern nutritional sciences (Mozaffarian et al., 2018).

The first half of the 20th century was an era of vitamin research and discovery (Carpenter, 2003). Until vitamin discovery, nutritional quality was solely measured by the intake of energy that food provided. Thiamine, isolated in 1926 and synthesized as vitamin B1 in 1936, was the first discovery of its kind (Mozaffarian et al., 2018). Casimir Funk identified thiamine when observing how the hulk of unprocessed rice protected chickens from beriberi. With the identification and synthesis of the essential vitamins and minerals, scientists studied their usages in preventing and treating nutritional deficiency diseases such as scurvy, beriberi, and pellagra. All of the major vitamins were isolated and synthesized before the mid-20th century, leading to dietary strategies in place for protection against scurvy (vitamin C), beriberi (vitamin B1), pellagra (vitamin B3), and other vitamin deficiency-related diseases. The chemical synthesis of vitamins also led to the vitamin supplement industry; multivitamins and bottled vitamin supplements were marketed and sold to individuals. Flour enrichment in the 1940s reintroduced vitamins and minerals lost during the refinement process (Huebbe & Rimbach, 2020). Food fortification with calcium, phosphorus, iron, and other specific vitamins became common. However, these fortifications had questionable benefits. For example, fortification on beer and unhealthy snacks allowed for marketing to claim falsified health benefits. National authorities in North America and Europe soon banned such marketing techniques. Authorities also announced the first recommended dietary allowances (RDAs) in 1941 at the National Nutrition Conference on Defence (Mozaffarian et al., 2018). Historical events such as the Great Depression and World War II ascertained precedent for nutritional-medical research and food policy recommendations.

The purpose of food processing changed in the late-20th century (Baker et al., 2020). Convenience and palatability were favoured over safety and nutritional value. Economic prosperity after the World Wars, urbanization, and increase of women in the workforce resulted in this increased demand for convenience (Welch & Mitchell, 2000). The food industry also altered; production from the producer was separated from industrial and retail processing. Ultra-processed foods (UPFs) became a substantial component of nutrition in high income countries (HICs) such as Canada, the United States, the United Kingdom, and Australia. Unprocessed elements of the 19th century diet (animal fats, fresh fruits and vegetables) were replaced by food items with increasingly processed components (seed oils, ready-to-eat snacks, to-go meals). Foods started to become associated with non-communicable diseases (NCDs) in HICs (Lee et al., 2022). Research at this time linked some UPFs to increase glycaemic load, reduce gut-brain signalling patterns, increase intestinal permeability and inflammation, and disrupt the endocrine system. A sharp increase in NCDs resulted in novel research studies and nutritional research programs.

The focus of nutritional studies from the 1950s to 1970s were on dietary fats, sugars, and protein (Mozaffarian et al., 2018). Studies comparing large groups of people rather than individuals, termed ecological studies, were typical. Short-term interventions in which one aspect of an established baseline would be changed in order to study the outcome were also common. However, ecological studies and short-term interventions use the nutritional model. The purpose of the nutritional model is to prevent deficiency diseases by isolating a singular nutrient and performing quantitative assessments of its effect and optimal intake. As a result, it is now known that the nutritional model is ineffective in preventing NCDs. From the 1950s to 1970s, ecological studies conducted by Ancel Keyes, Frederick Stare, and Mark Hegsted believed fat to cause heart disease. John Yudkin believed excess sugar contributed to coronary disease, hypertriglyceridemia, cancer, and dental issues. Contradictory studies were conducted by the Framingham Heart Study in the 1960s, where no link was identified between fat consumption and heart disease. By 1970, a growing body of research mainly using nutritional models determined over-consumption of food components including dietary fats and sugars to increase the risk of chronic diseases. Dietary recommendations in HICs shifted from ensuring adequate nutrient-intake to avoidance of excess intake (Davis & Saltos, n.d.). For instance, the 1977 Dietary Goals for the United States recommended a low fat and low cholesterol diet. During this period, reduced daily physical activity in the American population was also noted (Popkin et al., 2012). In lower-income countries (LICs), nutritional policies remained on ensuring adequate nutrient-intake and calories (Sukhatme, 2009). Protein deficiency was rampant among children in LICs, leading to the development of protein-enriched foods. The United Nations report on International Action to Avert the Impending Protein Crisis urged countries globally to increase the production of edible proteins. In response, the UK committee on nutrition aid and various other experts stated that the correct response to malnutrition should be through alleviating poverty instead of marketing processed, nutrient-enriched foods (Mozaffarian et al., 2018). Nonetheless, multinational industries promoted protein and nutrient-enriched foods, particularly baby formula in LICs.

Due to food processing techniques alongside economic development, malnutrition and nutrient deficiency were reduced in both HICs and LICs from the 1970s to the 1990s (Mozaffarian et al., 2018). During this period, rates of coronary mortality also decreased. Rates of chronic diseases such as obesity, type 2 diabetes, and various cancers, on the contrary, increased. With dietary guidelines focused on avoidance of excess intake, the food industry shifted; low-fat, low-cholesterol, reduced saturated-fat, and "lite" products were produced. Mainly in HICs, health and disease prevention of NCDs were placed on the consumer radar (Symbols et al., 2010). In 1973, nutrition labelling was proposed as a voluntary addition to food packaging in the United States by the Food and Drug Administration (FDA). Nutrition labels included the number of calories, grams of protein, grams of carbohydrate, grams of fat, the percentage of each macronutrient in relation to FDA guidelines, and the percentage of vitamin A, vitamin C, thiamin, riboflavin, niacin, calcium, and iron. It was made optional to include levels of sodium, saturated fat, and polyunsaturated fat. Mandatory nutrition labelling of products was implemented in July 1990 by the FDA on packaged and processed foods.

From the 1990s, scientific studies advanced remarkably; large multidisciplinary, multivariable, nutritional studies and randomized trials occurred (Mozaffarian et al., 2018). Cohort studies allowed for multivariable findings on nutrient and health outcomes at an individual level. Clinical trials were often used in high-risk areas to answer population-targeted questions. Genetic consortiums determined how DNA sequences influenced dietary choices, disease risk factors, and nutritional biomarkers. Due to this advancement, consistent evidence was provided to substantiate diet patterns and their effect on health.

Trials determined that previous dietary guidelines were ineffective; nutrient-based suggestions for the number of specific foods that should and should not be consumed did not result in positive health outcomes. Canada's 1992 revision of the Food Guide was historic, integrating the knowledge from many nutritional research experts (Canada, 2002). For one, the design incorporated a rainbow graphic displaying new names for the food groups: grain products, vegetables and fruit, milk products, and milk and alternatives. Most notably, there was a shift to the total diet approach which recognized that energy needs vary in individuals. The number of servings from each of the food groups represented a range in order to represent different age groups, body sizes, genders, and physical activity levels. Regardless of these changes, dietary guidelines delivered, and continue to deliver nutrient-based messages as opposed to messages about processing levels in food (Koios et al., 2022). The addition of the world's first insecticide-producing crop and herbicide-resistant crop through genetic engineering in 1992 by Flavr Savr tomatoes also revolutionized food processing (Rangel, 2015). The approval of genetically modified foods led to a tastier, increased supply of food with a larger shelf life. Tariff reductions under the North American Free Trade Agreement in 1994 also led to a rise in UPF food supply in Mexico and Canada (Baker et al., 2020). By the mid-2000s, there were four companies that dominated the UPF food sector. The four companies (Archer Daniels Midland, Bunge, Cargill, and Louis Dreyfus) controlled up to 90% of the global grain trade alone at that time. They were termed the "ABCD" companies as a result. Manufacturers heavily invested in equipment that would improve the taste and aroma of food through high-temperature processing, extraction, and fractionation. Packaging technologies also enabled novel food categories such as microwaveable popcorn and cake. It was during this time in the 1990s and 2000s when several notable psychologists gained interest on the link between nutrition and cognitive health (Clay, 2017).

NOTABLE STUDIES IN COGNITIVE SCIENCE

The earliest studies linking nutrition and cognitive health were on malnutrition, closely aligning with studies in the overall field of nutritional science (Edwards, 1979). Yoganada described two cases of individuals who lived without ingesting food for prolonged periods of time in a famous 1971 study. At that time, evidence linked semi-starvation and

14

starvation to severe impairment of cognitive function, emotion, and behaviour. The aforementioned shift in nutritional studies around the 1970s also extended to studies in cognitive science; poor diet due to certain foods, including UPFs, was associated with various cognitive diagnoses. Notably, a 1974 study by Greden found excessive coffee intake linked to cases of anxiety neurosis. Additionally, a 1978 study by Finn and Cohen found excessive coffee intake linked to severe migraines and body pain. Finn and Cohen also found agoraphobia and cardiac neurosis to result from coffee, tea, and tomato withdrawal. Schizophrenia was associated with the withdrawal of certain foods including coffee, eggs, and tobacco in a 1975 study by Lilliston. Alternatively, schizophrenia was connected to a dependency on niacin (vitamin B3) by Rosenberg in 1975.

In the past decade, research on the link between nutrition and cognitive health advanced exponentially (Clay, 2017). Researchers at Linyi People's Hospital in China suggested dietary patterns contribute to depression after conducting a meta-analysis of 21 studies from 10 countries (Li et al., 2017). A Western diet was associated with a smaller hippocampus in a multi-longitudinal study led by researchers at the Food and Mood Centre in Deakin University, Australia (Jacka et al., 2015). On the other hand, lower adherence to the Mediterranean diet was associated with attention-deficit/hyperactivity disorder (ADHD) in a study conducted by the University of Barcelona's Department of Nutrition, Food Science, and Gastronomy (Ríos-Hernández et al., 2017). UPFs were considered a factor of cognitive decline in adults, particularly those with type-2 diabetes in a new study published by the Alzheimer's Association (Weinstein et al., 2021). These studies support the growing field of nutritional psychology, offering potential prevention strategies and treatments to those who have been diagnosed with psychological conditions (Clay, 2017).

STUDYING UPF TRENDS OVER TIME

Understanding the changes in diet and UPF consumption over time is essential to take public health measures against NCDs. Dietary data from 1800 to 2019 in the United States were analyzed in one systematic review from 2022, highlighting trends in UPF consumption (Lee et al., 2022). In the review, total caloric and macronutrient intake per day were graphed (Figure 1).

The availability of UPFs for consumption is another important statistic to consider. For example, in 2016, Canada reported per capita UPF sales at 275 kg (Polsky et al., 2020). Mass-produced and packaged breads were the top UPF contributors. In fact, they were also the top contributor of UPF availability in the United Kingdom, Australia, and the United States. The availability of various fats and oils (Figure 2) alongside the availability of caloric sweeteners Figure 3) were graphed for the United States for a century (Lee et al., 2022). Overall, the levels of UPF intake ranged; 42% of total energy intake were from UPFs in Australia, with the United Kingdom and the United States having an average UPF intake of 57% and 58% respectively (Polsky et al., 2020).

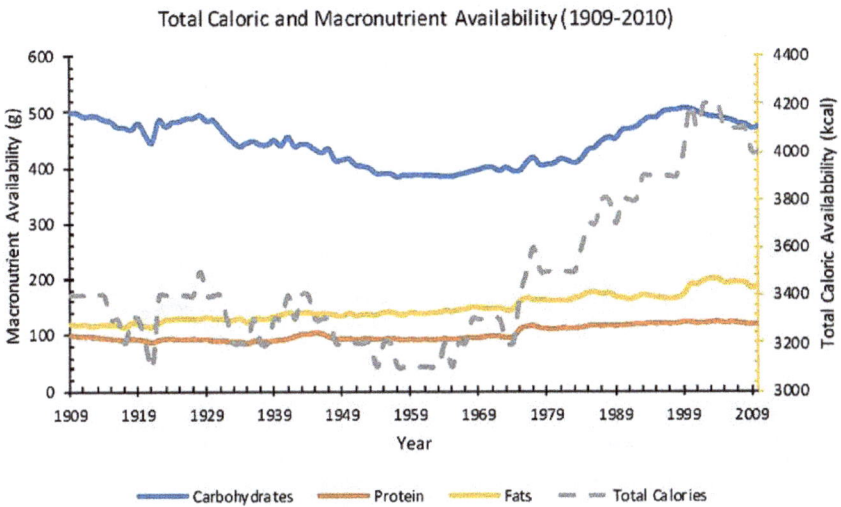

Total Caloric and Macronutrient Availability (1909-2010)

Figure 1: Total caloric, carbohydrate, protein, and fat intake per capita per day from 1909 to 2009. Total calorie intake increased 18% over the period, carbohydrate intake decreased 5%, protein intake increased 173% and fat intake increased 60% (Lee et al., 2022).

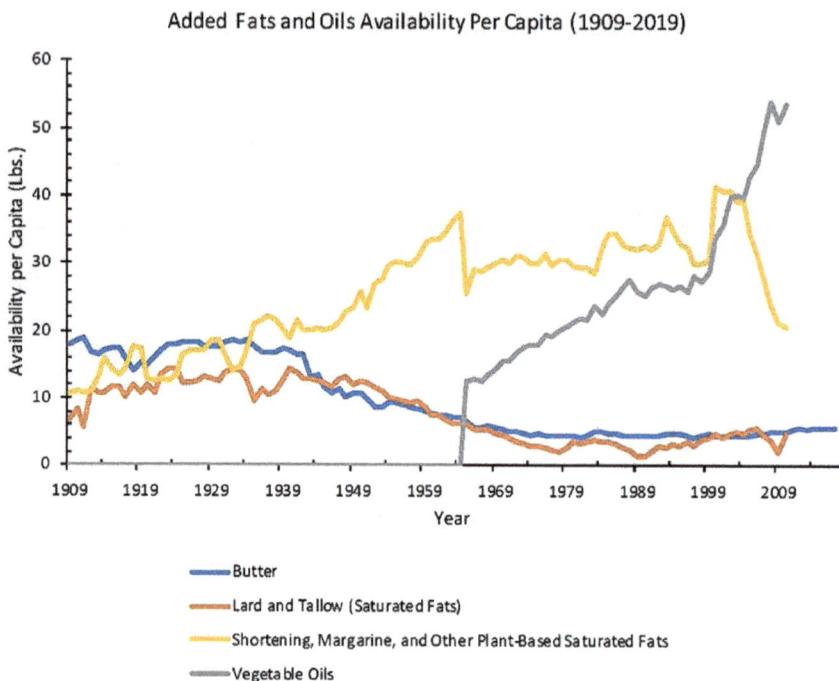

Added Fats and Oils Availability Per Capita (1909-2019)

Figure 2: The availability, per capita in lbs, of various fats and oils from 1909 to 2009 in the United States. The total availability of animal fats such as butter, lard, and tallow decreased by 58% in the century. The availability of butter decreased by 68% whereas lard availability decreased by 78%.On the other hand, margarine availability increased 192%, shortening availability increased 91%, and vegetable oil increased 329% (Lee et al., 2022).

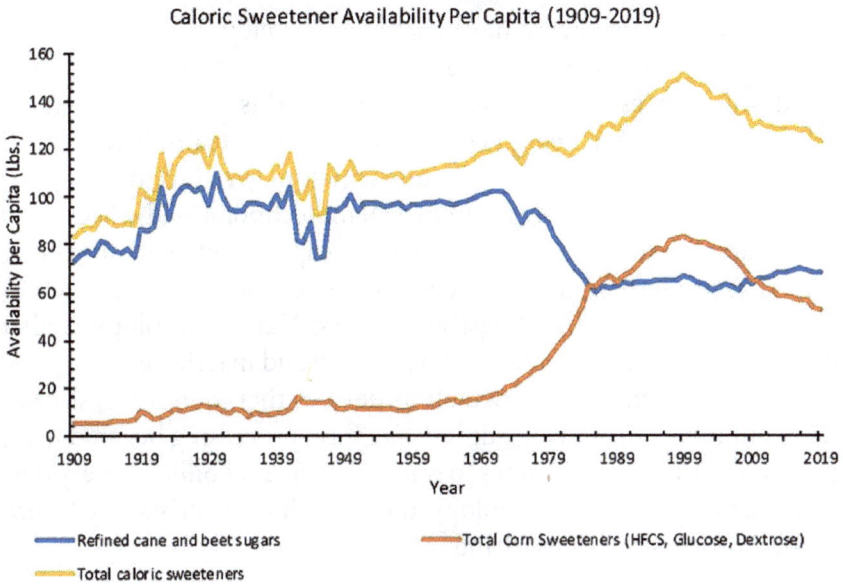

Caloric Sweetener Availability Per Capita (1909-2019)

Figure 3: The availability, per capita in lbs, of various caloric sweeteners from 1909 to 2009 in the United States. Overall, total caloric sweetener availability increased by approximately 200% in the century.

THE FUTURE OF FOOD PROCESSING

Nutrition scientists are currently working to eliminate poverty-related malnutrition while also trying to decrease the risks of NCDs due to the rise of UPFs (Weaver et al., 2014). There are emerging technologies that, when combined with advances in clinical, genetic, and metabolic medicine, have the ability to accomplish such a task. For instance, calorie overconsumption is the greatest contributor to obesity. To combat this, scientists are developing modified starches that reduce the rate of digestion in order to maintain a sense of "fullness." Noncaloric sweeteners such as Stevia have also been developed and marketed. Scientists have also tried altering the chemical structure of salt during the drying process to allow increased taste. This would help reduce salt intake. Steps are being taken to enhance the health benefits and safety of food. Bioactive compounds which may reduce disease risk and treat NCDs are being researched. Nanotechnology advancements have helped develop microemulsions that capture nutrients. The hope is to design microemulsions to control the release (or lack of release) of nutrients during the digestive process. Nanotechnology is also developing tools to improve food packaging and inactivate pathogens. This includes sensors and detection devices that identify freshness, mishandling, or water exposure. While food processing began during prehistoric times, it continues to impact the lives of billions every day. With the help of nanotechnology and nutrition scientists, the future of food processing is promising.

3

THE SCIENCE OF COGNITIVE DECLINE WITH ULTRA-PROCESSED FOODS

By Maureen Saha

INTRODUCTION

In today's world, ultra-processed foods (UPF) are a constant presence in the Western diet. In this fast-paced society, quick and easy meals are considered essential. UPFs are typically found in quick meals (foods that one can heat up and eat) and packaged foods that are made by companies, containing little or no whole foods, along with including flavoring, colours and other additives. It must be stated that most foods are processed in some form. Some foods need to be processed in order to make the item safe for consumption i.e milk. Other items contain whole foods, processed in some way for convenience, or to preserve their freshness - to contain them in cans, items such as canned tomatoes. However, there are some items that are heavily processed (ultra-processed foods), containing no whole foods, and plenty of chemicals, which will be the focus of this chapter. Easy to consume in excess, UPFs do not tend to make one full compared to whole foods like fruits and vegetables. Containing virtually no healthy nutrients, and usually high in sugar, salt and fat, ultra-processed food promotes inflammation, which is the biggest threat to healthy aging in the body and brain, along with being one of the primary reasons for cognitive decline (LaMotte, 2022).

THE HARM THAT ULTRA-PROCESSED FOODS CAUSE

A study that was recently conducted indicated that eating ultra-processed foods for more than 20% of one's daily caloric intake increases the chances of cognitive decline, affecting the areas of the brain involved in functioning, and the ability to process information and make decisions (LaMotte, 2022). Increased and continuous consumption also increases the risk of obesity, health problems, diabetes, and cancer. Consuming high levels of UPFs may disrupt the microbiome and the gut-brain axis, which may lead to neurodegenerative diseases. This, however, can be offset by incorporating plant fibre into one's diet, which is essential for maintaining the health and balance of bacteria in the gut, contributing to the gut-brain axis, and preventing the onset of neurodegenerative diseases (LaMotte, 2022).

Ultra-processed foods, typically branded as quick meals, contain virtually no nutrients that the body requires to function and ensure continuous processing. In some cases, manufacturers add synthetic vitamins and minerals to replace nutrients lost during processing. However, this is not a replacement for whole foods and the full range of essential nutrients that this provides. Replacing food that is high in plant fibre, quick meals increase the risk of age-related brain diseases like Alzheimer's. These quick meals typically have added sugar to them, which may lead to compulsive overeating. Additionally, due to the loss of fibre in the items during processing, it becomes easier for ultra-processed foods to be consumed, making it easier to eat more of these items in a shorter time period, along with consuming more calories. The presence of added chemicals for longer shelf-lives, and to make the item more palatable, may lead to adverse effects on the body. Albeit convenient for busy lifestyles, ultra-processed food leads to a host of issues in one's body. It is imperative that individuals are aware of the problems that ultra-processed foods may lead to, and realize that food choices that they make have an impact on their health today and in the future.

HOW NUTRIENTS GET INTO OUR BODY AND THE IMPACT THAT CONSUMING UPFS HAS ON THE BODY

The key player in one's body in breaking down nutrients is the digestive system, made up of the gastrointestinal tract, the liver, pancreas and gallbladder. The digestive system breaks down nutrients in the food that is being consumed into small parts, so the body can absorb and use them for energy, growth, and repairs inside the cells. Examples of nutrients include proteins, fats, carbohydrates, vitamins, minerals and water. Typically, proteins are broken down into amino acids, fats are broken into fatty acids and glycerol, and carbohydrates break down into simple sugars. The broken-down parts of the larger nutrients are typically absorbed by the small intestine and passed off by the circulatory system to other parts of the body to store or use. The liver is the place where nutrients are stored, processed, and delivered to the rest of the body when needed. Nutrients in foods that are highly processed tend to be delivered to the bloodstream right away, instead of being broken down and released slowly as the stomach and intestine digest them.

Processed foods typically contain high quantities of preservatives, saturated fats, salts, and low levels of nutrition and fibre. A diet that is high in fat is more prone to slowing down the digestive process, as lipids are harder for the body to break down (Fox, 2022). Increased salt intake leads to an increased chance of dehydrating the digestive tract, which may result in constipation. Fibre, a key nutrient in maintaining the digestive system, helps to keep bowel movements regular. Typically found in low levels in processed foods, this may lead to multiple issues involving the gut. Research has shown that high consumption of ultra-processed foods was associated with a higher risk of IBD (inflammatory bowel disease). The combination of low fibre, and high sugar causes the gut to be inflamed and makes it more prone to digestive disorders (Capetta, 2021). Making sure that individuals are consuming items that contain nutrients that positively contribute to gut health is of utmost importance, as this is the beginning of the process of nutrients being circulated to the rest of the body.

WHAT'S IN PROCESSED FOODS THAT MAKE US LIKE THEM?

Ultra-processed foods are typically something that is addicting and easy to overconsume. Typically calorie dense, it is quite easy to continue to eat and consumes more calories than we realize. Research has shown that sugar can register in our brains in less than a second (Link, 2021). Processed foods, typically high in sugar, easily trigger our desire and get into our stomachs much faster. Processed foods tend to be more calorie-dense (contain more energy) compared to non-processed, thus they deliver food more quickly. Speed is the important component that results in the addictive nature of processed foods. Diets that are ultra-processed tend to have foods that are more calorie dense, less satiating, and more satisfying. Thus, to achieve the same level of fullness as eating non-processed foods, an individual on an ultra-processed food diet would have to eat more of the calorie-dense food, consequently accumulating additional calories (Link, 2021).

Evolutionarily, our bodies are designed to get and find calories as much as we can, because the future is unpredictable; it is better to store energy at the present moment than to wait for the unknown future (Link, 2021). Processed foods operate in a way that utilizes the evolutionary mechanism to their advantage- foods that are calorie-heavy that cause the brain to overeat. Research conducted at the U.S National Institute of Health found that individuals tended to eat more ultra-processed food, and they ate it faster- indicating a link between ultra-processed food and weight gain, among the plethora of health problems (Young, 2019). This research clearly demonstrates the link between ultra-processed foods and the growing obesity epidemic. Aggressive promotion, labelling, bright colours, and widespread advertisement also play a role in enticing consumers. Many times, consumers may not realize how many calories or sugar is in the item that they are consuming, highlighting the importance of making sure to check the nutrition labels and the ingredient list of all items.

KEY NUTRIENTS THAT WE NEED AND WHY

It is essential for the body to consume nutrients in order to properly function and repair cells. Ultra-processed foods, dense in calories, and containing very few nutrients are associated with neuroinflammation and a reduction in cognitive function. During the manufacturing process, nutrients that are found in unprocessed whole foods get altered and removed, leaving behind an item that is not as nutritious. A health-healthy lifestyle promotes a healthier brain, along with helping to improve cognitive function. A heart-healthy lifestyle consists of a diet rich in whole grains, fresh fruits and vegetables, and lean protein, along with adequate sleep and exercise (Rapaport et al., 2022). Individual nutrients, such as vitamins and polyunsaturated fatty acids play an important role in the enhancement of cognitive performance (Klimova et al., 2001). In one's diet, fats and proteins also play an important role in maintaining brain function. Research has found that the amino acids in protein after it breaks down are used by the body to produce neurotransmitters (Tremblay, 2018).

Healthy fats, such as Omega-3 fatty acids, found in fish and some nut and seed oils work to promote brain function. Low levels of these fats in one's body can cause neurological disorders and cognitive disorders (Tremblay, 2018). Vitamins, such as B-12, help to keep brain cells coated with myelin, which is a fatty substance that helps the nerves to communicate effectively, along with maintaining synaptic function plasticity (Tremblay, 2018). On the other hand, low dietary intake of omega-3 polyunsaturated fatty acids can contribute to depression through its effects on inflammatory pathways in specific brain regions, along with contributing to memory loss (Tremblay, 2018). Increased consumption of fruits and vegetables that are high in polyphenolics (organic compounds that help manage blood pressure and promote good circulation) can prevent, and in some cases reverse age-related cognitive deficits, lowering inflammation and stress (Tremblay, 2018). In order to ensure that individuals are consuming the essential nutrients, a person needs to be conscious of the items that they are picking up at the grocery store and ensure that they choose whole foods or minimally processed items.

CURRENT TRENDS SURROUNDING UPFS IN THE POPULATION

Food and nutrition have always been essential in the maintenance of brain performance. One's overall composition of the human diet and the specific dietary components that make it up have massive impacts on brain function (Ekstrand et al., 2020). The effects of nutrition take place before birth. The diet of a pregnant mother and the nutritional status of a newborn baby and young infant influences the development of the brain and influences future development (Ekstrand et al., 2020). Lack of nutrition affects neurodevelopmental processes, affecting processes such as neuron proliferation and myelination (Prado & Dewey, 2014). However, easy access to ultra-processed foods for young children and adolescents has an impact on critical periods of development. Research has found that ultra-processed products are at the forefront of food supplies found in high-income countries, along with the consumption of ultra-processed foods increasing in middle-income countries (Edwin Kwong Research Fellow et al., 2022).

Research has discussed that ultra-processed foods are unlikely to increase further in high-income countries, such as the US and Canada (Edwin Kwong Research Fellow et al., 2022). However, currently, middle-income countries such as Brazil and Indonesia are being targeted by big companies. A recent study has found that the total amount of ultra-processed foods that are bought in middle-income countries will be almost the same as in rich countries by 2024 (Edwin Kwong Research Fellow et al., 2022). Although there are fewer people going hungry- fewer people suffering and dying from insufficient food, their individual diets have not gotten better. Subsequently, there is a rapid rise in the rates of global obesity, along with rising rates of diet-related health problems globally. The increasing availability and affordability of unhealthy foods are one of the primary reasons that obesity could be a global issue. Making sure to have positive discussions surrounding the importance of choosing healthier options is vital in ensuring that individuals are cultivating good habits regarding their diet.

Struggle that the population faces in accessing good food
Although there is a constant conversation surrounding the need to change diets and incorporate healthier foods into our diet, it must also be acknowledged that this may not be possible for everyone. Due to the increase in availability and affordability of fast food/ ultra-processed foods, their consumption is increasing in the population. A number of socioeconomic factors, such as low access to healthy foods, less time to prepare foods from scratch, and being unable to afford whole food options, are leading families to resort

to incorporating processed foods as a major part of their diet. Families with parents who work multiple jobs, are short on time and need to choose between feeding their families or taking care of other expenses. Economically, the price of fruits and vegetables is increasing, while the price of processed foods is going down. Rapid urbanization is a crucial component of their changes- farmers are choosing to give up farming and move to the cities, along with individuals in cities resorting to processed foods as it is not time-consuming, and more economical. Although one may think that it is an individual issue, the reality is that the unaffordability of healthy food is a systemic issue, which requires governments to step in and enact policies and programs which would make it easier for families in low-income families to be able to access healthier foods.

WHAT CAN BE DONE, AND NEXT STEPS

Affordability, along with lack of knowledge, is the barriers that are currently in place preventing individuals from accessing healthier food options. Policies need to be enacted taking into consideration individuals' incomes and making affordability a priority. Currently, processed foods are better priced than healthy foods, which in the long run, will lead to an increase in diseases in the population, along with more strain on the system. Additionally, the implementation of programs discussing the health benefits of healthy foods should ideally be incorporated into the education curriculum. Rather than discussing and simply stating that students should be making healthy food choices, it would be ideal to go further into a discussion about the importance of consuming the nutrients that are present in healthy foods, and the impact that this may have on the body. It is important to discuss, however, that most of the food that we consume is processed in some way.

It is important to choose items that are minimally processed, and limit the consumption of ultra-processed food from the diet. Emphasizing the importance of checking nutrition levels, along with making sure to double-check ingredients, will ensure that students are more conscious at the grocery store. Checking the nutrition label will allow the shopper to be aware of the levels of hidden sugars that may be added to the item. On the contrary, leading discussions surrounding ultra-processed foods, and their lack of nutrients, along with the impact that this can have on the body, may lead to more awareness when consuming and choosing items to consume. Education, along with enacting policies and programs to make healthier foods accessible for everyone is the only way that the population will turn out healthier, and happier.

4

THE ECONOMICS OF THE ULTRA-PROCESSED FOOD INDUSTRY

By Wan Ling Dai

INTRODUCTION

The rise of ultra-processed foods came about during the 20th century as science and technology began being applied in agriculture, and food and beverage manufacturing (Capozzi et al., 2021). This led to an industrialization of food production systems and the idea of ultra-processed foods. Food processing has increased the quality of taste, shelf-life, and safety of food products. As a result, consumers transitioned from a traditional diet consisting of home-cooked meals to commercially prepared meals made with processed foods. However, the health potential and environmental footprint of processed foods are becoming a growing concern.

FOOD PROCESSING INDUSTRY

The majority of ultra-processed foods available for consumption are made to have a long shelf life and made ready-to-eat, such as breakfast cereals, crackers, and granola bars. Currently, ultra-pro-

29

cessed foods dominate the global food supply and provide more than 50% of dietary calories of an individual person (Capozzi et al., 2021). The food processing industry is primarily driven by the goal to increase profits, which is usually achieved by minimising costs across all stages of production. This is often done by using cheaper alternatives of food ingredients wherever possible. For instance, solid margarines were created by converting the unsaturated fatty acids in vegetable oils to saturated fatty acids to make the oil more solid (Capozzi et al., 2021). The solid margarines were then marketed as cheaper and healthier than butter. It was later found that the high content of trans fatty acids found in foods, such as the margarines, had severe effects on the cardiovascular health of individuals, resulting in trans fat being restricted or banned in several countries. This showcases the potential negative effects of food processing. Despite that, food processing can also improve the nutritional value of food products by providing essential nutrients.

The food and beverage processing industry is the second largest manufacturing industry in Canada with it accounting for 17 percent of total manufacturing sales and two per cent of the national Gross Domestic Product (GDP) (Government of Canada, 2021). This industry also supplies approximately 70 per cent of all processed food and beverage products available in Canada (Government of Canada, 2021). The remaining processed food and beverage products are exported to various countries with the majority being supplied to the United States, China, and Japan. The export of these processed products generated a revenue of 38.9 billion dollars in 2019 with an average annual growth rate of 6.9 per cent (Government of Canada, 2021). The largest food and beverage processing sector is the meat product manufacturing which accounted for 25 per cent of all manufacturing sales, valuing at 30 billion dollars, in 2019 (Government of Canada, 2021). The meat product manufacturing sector is also the most significant food industry in Quebec, Ontario, Manitoba, and Alberta (Government of Canada, 2021). The food and beverage processing industry is also the largest manufacturing employer in Canada with 7,800 processing establishments and 290,000 employees (Government of Canada, 2021).

TRENDS AND OPPORTUNITIES FOR FOOD INNOVATION IN CANADA

Agriculture and Agri-Food Canada (AAFC) identified trends in conumer preferences and market pressures, and opportunities in innovative ingredients and emerging technologies for the food processing industry (Government of Canada, 2015). AAFC determined that the factors that influence customer preferences are shifting demographics, convenience, environmental stewardship, and desire for more information about food (Government of Canada, 2015). The shift in demographic is a result of ageing Baby Boomers, the growing purchasing power of Millennials, and the increasing ethic diversity of the Canadian population. Millennials tend to seek out information and are more focused on environmentally friendly practices. Millennials also tend to be more health conscious. This new demographic has resulted in changes in customer preferences to more nutritious, and ethical and sustainable foods. As the global population is expected to reach 9.7 billion by the year 2050, the food and beverage industry is challenged to produce more to meet the demand of the growing population (Government of Canada, 2015). This places a further desire to improve the sustainability of food production. Thus, there is an opportunity to create innovative ingredients to help meet consumer and marketplace pressures. There are three approaches to creating innovative ingredients: biofortification, fortification, and supplemented foods (Government of Canada, 2015). Biofortification is the process of increasing the nutritional value of plants or animals through selective breeding, generic engineering or adjusting animal feed (Government of Canada, 2015). Fortification is the addition of vitamins and minerals to staple foods to restore nutrients lost in processing (Government of Canada, 2015). Supplemented foods are products that have added substances, such as vitamins, minerals, and amino acids, with the intent of providing a health benefit (Government of Canada, 2015). Emerging technologies include opportunities that arise from new knowledge and the innovation application of existing knowledge. Some of these new technologies are biotechnology, nanotechnology, packaging, and applied technologies (Government of Canada, 2015).

PRODUCERS OF PROCESSED FOODS

The top ten largest publicly traded food companies in 2019 are Nestle, PepsiCo, Anheuser-Busch InBev, JBS Foods, Tyson Foods, Archer Daniels Midland (ADM) Company, Mars Wrigley, Cargill, The Coca-Cola Company, and Kraft Heniz Company (Food Engineering, n.d.). Nestle is the largest food and beverage company in the world with sales of 80.195 billion USD (Food Engineering, n.d.). Nestle produces a large variety of food products, ranging from baby food to frozen foods to confectionery. PepsiCo generated a revenue of 64.661 billion USD in 2019 (Food Engineering, n.d.). PepsiCo also sells a large variety of food products, but they are most known for their PepsiCo beverages, Frito-Lays potato chips, and Quakers Oats Foods. Anheuser-Busch InBev had sales of 54.619 billion USD and they are the world's largest beer brewer (Food Engineering, n.d.). JBS Foods had sales of 46.79 billion USD in 2019 and they produce a diverse selection of protein products for consumers around the world (Food Engineering, n.d.). The fifth largest food company is Tyson Foods with a generated 40.052 billion USD in revenue in 2019. Tyson Foods is the world's second largest processor of chicken, beef, and pork after JBS. Archer Daniels Midland (ADM) Company generated sales of 38.9 billion dollars in 2019 (Food Engineering, n.d.). ADM Company operates agriculture services to take natural products and convert them to renewable industrial products, and food and beverage ingredients and solutions. Mars Wrigley had a revenue of 33 billion USD in 2019 and they are the world's leading manufacturer of chocolate, chewing gum, and other confections with some of their most known products being M&M's and Skittles (Food Engineering, n.d.). Cargill had sales of 32.5 billion USD (Food Engineering, n.d.). Cargill provides food, agriculture, financial, and industrial products and services to countries across the globe. The Coca-Cola Company generated a revenue of 31.856 billion dollars in 2019 and they are a multinational beverage corporation, with their most known product being the sugary drink, Coca-Cola (Food Engineering, n.d.). The tenth largest food company is Kraft Heniz Company with sales of 26.259 billion USD in 2019 and they are a global producer of foods, with their most popular products including their Heniz Ketchup and Kraft Mac & Cheese.

DEMAND FOR PROCESSED FOODS

The global ultra-processed food industry is projected to grow at a compound annual growth rate of 6.01 per cent with a forecasted value of 2,041.271 billion USD by 2025 (Processed food market - forecasts from 2020 to 2025, n.d.). This rapid growth is driven by the increasingly busy lifestyles and the reduction of available leisure hours of consumers. Moreover, this has led consumers to seek food products that are quick and convenient, and thus, significantly driving up the demand for ultra-processed foods. Additionally, the growth of e-commerce has allowed consumers to easily access ready-to-eat and processed foods compared to perishable items.

CIRCULAR ECONOMY OF PROCESS FOODS

A circular economy is an economic system that attempts to recover as much as possible from resources by reusing, repairing, remanufacturing, repurposing, repairing, or recycling products and materials to ensure nothing is wasted. This promotes innovative ways to improve the environment and economy. With the rise in demand for sustainable and healthy foods, processing technologies need to be developed to create advancements in a circular economy. The technologies must be more efficient with water consumption and responsibly utilise all natural resources in order to meet the demand for sustainable, safe, and nutritious foods.

In order to further progress the circular economy, there needs to be advancements in the sustainability of food supply chains. This also ensures global economic and social development, and sufficient resources are available for future generations. In order to do so, the three main effects of production processes in the food supply chain system need to be improved; these three main effects include environmental, economic, and social impact. These three effects can be improved by lowering the ecological footprint of the transformation of raw ingredients into food. Hence, the food loss and food waste produced between the distribution and consumption points in the food supply chain should be minimised. However, this is a difficult task as any type of food production generates waste as not all parts of raw ingredients are edible. For instance, fruits and vegetables produce a significant waste of 25 to 30 percent of all raw material waste as the seeds, skin, rind, and pomace of fruits and vegetables are often not consumed (Capozzi et al., 2021). One method of doing so is to integrate industrial sectors such that there is a clear pathway for the use of by-products of one sector to be used as the raw materials of another sector. Another method would be to produce nutritious foods using renewable raw materials as this makes the food supply chain more sustainable and lowers the social and environmental impacts. The benefit of this method is that it would continue to make food available at an affordable cost, which is especially important in the current rise of food prices.

AFFORDABILITY OF PROCESSED FOODS

One of the significant contributors to an individual's diet is the price and affordability of foods. Food processing facilities have the ability to produce processed foods more efficiently than farmers can grow crops. Moreover, processed foods are priced lower than fresh foods as commodity crops that are used to produce processed foods are often subsidised, whereas fruits and vegetables that are unprocessed or minimally processed must be privately funded (Larson, 2022).

IMPACT OF COVID-19 ON FOOD AND BEVERAGE COMPANIES

Due to the pandemic, many food and beverage companies are facing significant reduced consumption and disruptions in their supply chain systems. The COVID-19 crisis created a labour shortage, hence livestock production, and planting and harvesting of crops decreased (Aday & Aday, 2020). In addition, across the world, farmers are facing a shortage of commercial fertilisers (Chaarani, 2022). Hence, these farmers did not have enough supply to meet the demands of these food and beverage companies. This results in a butterfly effect as these companies are now forced to produce a lower quantity of products. These limited quantities of supply are in turn increasing the demand of these products and resulting in the rise of food prices. Additionally, due to the pandemic, production at these food and beverage plants were reduced, suspended or temporarily discontinued or workers were reluctant to work due to the fear of contracting COVID-19 so these companies are producing at a lower capacity than before the pandemic (Aday & Aday, 2020). For example, due to the aforementioned reasons, the production capacity of pork facilities were predicted to have decreased by approximately 25 per cent in late April of 2020 (Aday & Aday, 2020). This was especially a large problem with companies that used a small number of large production facilities as an entire facility would need to be closed in the event of an outbreak which reduces the company's production line and the company's ability to meet customer demands.

The presence of the pandemic also changed how customers spent their money on foods. Due to the boredom experienced in quarantine, some consumers began consuming high quantities of fat, carbohydrates and proteins, while others consumed sugary foods to feel positive as carbohydrate rich foods can encourage serotonin production (Aday & Aday, 2020). The closure of restaurants and limited eating services resulted in a shift from food service to food retail. Reports found that the purchasing from supermarkets and using food services had the same ratio at 50 per cent before the pandemic; however, the pandemic, consumers purchased from supermarkets at a rate of almost 100% (Aday & Aday, 2020). Consumers were also found to spend more money per visit to the grocery store during the pandemic and focused on purchasing products with longer shelf life, such as canned foods and frozen foods, due to the convenience of these foods. In a study conducted in European countries, it was found that the demand for frozen vegetables increased by 52 per cent in the week that the COVID-19 pandemic was announced (Aday & Aday, 2020).

STRATEGIES FOR IMPROVING FOOD SUPPLY CHAIN

Prior to the COVID-19 pandemic, one-third of all food produced for human consumption was wasted across all stages of the food supply chain. The presence of the COVID-19 pandemic highlighted the need to reduce waste as it challenged companies to compete and grow in an environment where resources were scarce. One method of doing so is to produce valuable bioactive components, such as phenols, carotenoids, pectins, flavonoids, essential oils, glucosinolates, isothiocyanates, and whey protein isolate using the food waste and reuse these components in the food supply chain (Aday &

Aday, 2020). These compounds can be used as preservatives, gelling agents, food or nutritional supplements (Aday & Aday, 2020). This could in turn produce foods with a longer shelf life and more nutrition.

The COVID-19 pandemic also highlighted the difficulties of human resource management, which include the change of working conditions, adopting new workplace policies, and actions to minimise human contact (Aday & Aday, 2020). Protocols to wear face protection equipment, to perform temperature screening of staff, and to avoid overcrowding of workers are necessary for food and beverage companies as it will reduce the possibility of a COVID-19 outbreak, ensure employees feel comfortable going to work, and help maintain continuous production of goods.

Companies that used the centralised method of production, wherein there was a small number of large manufacturing facilities, should instead switch to the decentralisation of food manufacturing (Aday & Aday, 2020). The decentralisation of the production of processed foods allows for small facilities and hence, in the event of a COVID-19 outbreak at one plant, there is a smaller impact on the company's overall production capability. Another benefit is that these companies could open facilities close to consumers and reduce storage and transportation costs of products and minimise the environmental effects associated with the production of food products (Aday & Aday, 2020). In addition, these companies could open food processing plants closer to suppliers and agricultural inputs as this would also decrease storage and transportation costs to obtain raw ingredients.

Overall, there is a strong growth in the ultra-processed food industry. However, there is also a demand for healthier foods and for food and beverage companies to reduce their ecological footprint. Hence, food and beverage companies are facing the challenge to create sustainable foods that are highly nutritious while being more affordable than their fresh food alternatives.

5

THE POLITICS OF THE ULTRA-PROCESSED FOOD INDUSTRY

By Michael Tang

THE POLITICAL IMPACT OF THE ULTRA-PROCESSED FOOD INDUSTRY

Ultra-processed foods have become a cornerstone of the modern diet due to how they can be less expensive than healthier whole foods and their overall pleasurable taste due to inclusion of ingredients such as refined sugars and fats. The presence of processed foods is only bolstered further by the influence that many processed food companies have over the general population due to their aggressive advertising strategies and significant influence over the general food supply chain. While it is clear that reducing the production and consumption of ultra-processed foods and beverages would likely have a salubrious impact in almost any country around the world, these foods remain prevalent as ever with varying degrees of regulation across the world. One of the less obvious contributing factors to this situation is the industry's active participation in political activity and the significant influence that ultra-processed food companies and interest groups can hold over government officials, other influential figures (such as researchers), and society. Even though many countries have introduced and are continuing to introduce policies aimed at improving public health (such as soda taxes and nutrition programmes) and decreasing the prevalence and incidence of non-communicable diseases (NCDs) such as obesity, the ultra-processed food industry manages to hamper or manipulate these policies to their benefit. The ultra-processed food industry as a whole can leverage its economic impact and financial earnings to sway the policies enacted and enforced by local governments and cause social divisions or distractions that make it more difficult to create widespread, concerted social action against the industry's interference (Gómez, 2021).

Processed food companies often try to get involved with public health and governmental efforts aiming to address unhealthy eating and public health. This appears strange at first as policies aiming to promote healthy eating would generally harm the production, sales, and profits of processed foods. However, the involvement of these companies and their allies may serve to control or manipulate the implementation of policies and laws to their favour, likely leading to a more favourable outcome compared to what would have happened had they not gotten involved in the first place. A common trend among public health policies targeting NCDs and nutrition is their focus on prevention rather than regulation, suggesting that governments would rather focus on individual choices rather than larger-scale policies that would more directly harm the ultra-processed food industry (Gómez, 2021). Some common prevention policies include improving nutritional education and food taxes to discourage the consumption of certain foods. Even regulatory public health policies take a relatively modest approach, often focusing on other factors such as nutritional subsidies, reformulating food, or regulating the marketing and sale of food products (Gómez, 2021). It makes sense why this would be the case, given the political and economic costs of antagonizing the ultra-processed food industry due to their influence in society. Unfortunately, this focus on prevention rather than stricter regulation often hampers the ability of governments to create a significantly healthier environment for its citizens and improve overall public health, even though prevention policies can still have positive impacts (Gómez, 2021).

38

INTEREST GROUPS, LOBBYING, AND FINANCIAL INCENTIVES

Similar to many other industries, processed food companies make use of interest groups and lobbying to attain their political objectives. Interest groups that are formed with political objectives in mind and put significant pressure on policymakers can be considered as "special interest" or "pressure" groups (Mariath & Martins, 2020). Members of pressure groups maintain contact with policymakers to try to influence them to enact policies in favour of the self-interests of a restricted economic or social group. Special interest groups are a subtype of interest groups that generally have more power and influence over policymakers than other groups due to their effective organization (Mariath & Martins, 2020). In the case of processed food interest groups, they are solely focused on promoting the interests of processed food companies and associates and can be highly organized and powerful due to their organizational support from the large and influential processed food industry. Regardless of their classification, these groups engage in lobbying activities to exert political influence.

Lobbying generally refers to any legal attempts made by individuals or groups to influence government policies or actions. The lobbying process involves careful information collection, the preparation of policy drafts, creating strategies to defend such drafts, and seeking political allies among other activities (Mariath & Martins, 2020). The exertion of pressure is usually only done as the final stage, after the foundational information and connections have been established. Additionally, lobbyists often have specialized political knowledge that allow them to create effective legislation and regulation drafts and the connections to help push those drafts through (Mariath & Martins, 2020). Interest groups supported by influential and highly organized groups or industries can hold significant advantages or privileges over other groups to the point where public interest may be compromised in some situations. Public and trade union interest groups often cannot compete with the financial resources that private sector interest groups have access to, leading to an inequality in the degree of participation and influence during the policymaking process (Mariath & Martins, 2020). Private sector interest groups simply have more resources to hire the most effective lobbyists (in terms of skill and pre-existing political connections) and likely hold more influence over politicians due to their economic impact on society. Just as industries are regulated by government policies, the government is dependent on the economic revenue and opportunities created by those industries. This leads to a reciprocal relationship where industries can control or influence the government to a certain degree.

The ultra-processed food industry can at times use their vast financial resources to directly or indirectly provide financial incentives for policymakers. Direct incentives include gifts, donations, and other forms of financial inducements while indirect incentives include the promise or potential of further economic opportunities (e.g., jobs) and revenues generated for the government via taxes (Moodie et al., 2021). The industry can also threaten legal action or other political inconveniences due to the significant financial and logistical resources they have (Moodie et al., 2021). Incentives and threats of inconveniences, when paired with lobbying activities and the formation of political connections, allows the ultra-processed food industry to have significant political influence and manipulate any policies that have the potential to affect its operations and profits.

CONSTITUENCY-BUILDING ACTIVITIES

Another method that the ultra-processed food industry employs involves constituency-building activities that cause divisions within society, preventing large-scale social opposition against the industry (Gómez, 2021). In general, these activities involve increasing the industry's involvement in society to gain the support of influential individuals and non-governmental organizations (NGOs), improve the industry's reputation through social marketing, and divert society's attention away from the harmfulness of their products. These supporters can then cause division or confusion in society (intentionally or unintentionally) as individuals and organizations supporting public health policies are pitted against individuals and organizations that support the processed food industry. Some of the key individuals targeted by the industry in their constituency-building activities include scientific researchers and civic advocates such as libertarian economists (Gómez, 2019). For example, the processed food industry may support poor communities

40

by providing research grants to local academic researchers and social programmes to gain the support of local NGOs and citizens (Gómez, 2019). The industry may also establish or support programmes and sports initiatives focused on promoting exercise as a way of improving their image as a socially conscious industry while simultaneously diverting attention from the relationship between processed foods and chronic diseases (Gómez, 2019). While lack of physical activity is certainly a contributing factor to many chronic diseases and addressing this issue would no doubt help to improve public health, the ulterior motives behind the industry's involvement is problematic. They are more likely to be promoting exercise as a way to improve their own image or distract everyone from the consequences of their own products rather than being motivated by purely philanthropic interests. Even though there are many reputable scientists and NGOs that work to reduce the impact of processed food in society, the presence of a strong counterpart serves to counteract their efforts. These constituents can potentially exert a more direct political effect through voting for or supporting politicians that are more lenient towards the processed food industry. Overall, these influential individuals and organizations act as the industry's spokespeople and representatives in society, thus allowing the industry to have a large social presence in an indirect form that does not draw as much attention to the industry itself.

Public health organizations and academia are a key target in particular as the support of researchers is often key in making governments disregard important evidence on topics such as obesity. The industry engages in activities such as funding particular studies, disseminating cherry-picked data, spreading unpublished or questionable evidence, directly criticizing unfavourable data, and hosting their own scientific events (Moodie et al., 2021). These actions all seek to add uncertainty or complexity to the scientific debate on topics such as obesity, allowing the government to have an excuse to potentially favour the industry on certain health-related policies. One example of the impact of industry-funded research groups is the International Life Sciences Institute (ILSI). The ILSI has helped the industry to be involved with scientific research in several countries, and has successfully lobbied the Chinese government to adjust its obesity policy (Moodie et al., 2021). As a result, the policy has shifted its focus to a lack of physical exercise rather than diet. Many public health organizations stressed the importance of the obe-sogenic environment caused in large part by ultra-processed foods, and how the industry should be more strictly regulated. Unfortunately, industry-funded researchers and evidence have served to undermine the credibility of anti-processed foods research and policies. Furthermore, the industry may also take more direct action against certain researchers by threatening legal action, monitoring their individual activities, and using the media to attack their reputation (Moodie et al., 2021). The industry can also poach advocates or researchers (using methods such as financial incentives) from anti-processed food groups to strengthen industry-funded groups while simultaneously weakening and impacting the reputation of anti-processed food groups. Overall, the ultra-processed food industry can utilize its immense resources to target researchers and public health groups that advocate for action against processed food products.

COCA-COLA IN MEXICO: A CASE STUDY

One notable example of the ultra-processed food industry's political activities is Coca-Cola's influence in Mexico. The situation in Mexico is of interest as the country has one of the highest numbers of obesity and type-2 diabetes cases in Latin America and a decentralized healthcare system where local governments have significant control over policymaking and financing relating to health administration (Gómez, 2021). Coca-Cola is the largest soda industry in the country, taking up around 70% of the total market shares of soft drink sales in Mexico in 2018 (Gómez, 2019). Coca-Cola has repeatedly been successful in preserving its ability to market and sell its products despite government action to prioritize NCD prevention, which should theoretically include tackling the consumption of soft drinks as studies have found that these beverages can increase the risk of conditions such as dementia and stroke (Anjum et al., 2018). Four of the major factors behind Coca-Cola's influence in Mexico include: its bureaucratic context, connections with policymakers and the ability for lobbyists to meet with such politicians, connections with the president, and a "conflictual civil societal response" caused by the conflict between Coca-Cola-funded researchers/activists and public health advocates (Gómez, 2019).

42

In countries with high concentrations of political and policy-making authority, politicians can be disinclined to work with civil interest groups and organizations due to factors such as a sense of elitism and a lack of faith in the general citizen's political ability (Gómez, 2021). This is the case with the Mexican government, with policymakers often unwilling to engage with activists promoting nutrition and anti-obesity policies. However, Mexican policymakers are more willing to engage with representatives of the processed food industry, whether it be due to the industry's importance in society or the potential for benefits (Gómez, 2021). This leads to a situation where industry-backed lobbyists often have more pre-existing connections with and the ability to access policymakers. Given the authority that policymakers hold and the interest groups that they choose to engage with more often, processed food companies such as Coca-Cola hold strong influence over policymakers the creation and enforcement of policies. Additionally, Mexican presidents have access to significant political powers, from the legislative veto authority to the ability to appoint and dismiss federal agency directors (Gómez, 2019). As such, past presidents with personal ties to the company and a vested interest in the success of the company are able to steer policies in their favour. For example, former Mexican President Vincente Fox Quesada (2000-2006) was a Coca-Cola executive before his term. He received substantial campaign donations from Coca-Cola and allowed the company to have significant influence over health policy (Gómez, 2019). For example, he refused to implement a tax on sodas that used sugar cane ingredients in 2003 under pressure from several industries including Coca-Cola. Other former presidents, such as Felip Calderón (2006-2012) and Enrique Peña Nieto (2012-2018), also supported the widespread sale and distribution on Coca-Cola products, with Nieto promoting Coca-Cola products (e.g., biscuits) as part of his anti-hunger programme (Gómez, 2019).

The "conflictual civil societal response" essentially refers to the impact of the constituency-building activities that Coca-Cola engages in. Coca-Cola has backed several researchers and organizations in Mexico, causing division in academia and in society surrounding the relationship between Coca-Cola's products and NCDs and what kind of policies are needed to address this issue (Gómez, 2019). As part of these activities, Coca-Cola has been active in promoting physical exercise as a solution to obesity and other chronic diseases in Mexico. The company has contributed to the establishment of international research organizations that raise awareness about the lack of physical exercise and claim that physical inactivity is responsible for chronic diseases more so than excess calories and ultra-processed food products. Organizations such as the Exercise is Medicine global foundation have branches in Mexico and are headed by well-known nutritional scientists or other influential figures (Gómez, 2019). Coca-Cola has also sponsored sports events and local and national soccer clubs to further improve its social marketing and influence.

DIRECTIONS MOVING FORWARD

Overall, processed food companies and interest groups thoroughly leverage their immense resources and influence over politicians and influential figures in society to ensure that policies are generally enacted either in their favour or in a way that does not negatively impact their operations. While it is certainly difficult to combat or contain the processed food industry's impact on governments and societies, an important first step is the widespread recognition of the actions and methods that the processed food industry uses to manipulate local politics. This enhanced awareness could encourage the general populace to hold local politicians more accountable regarding their nutritional policies and inspire further social protest or action against this industry and their supporters. The focus on public health prevention policies, nutritional education, and increasing access to healthier food options should continue as strict regulatory changes would not be very useful if citizens continue to seek out processed food options for reasons ranging from habit to necessity. However, it is also important for citizens, NGOs, and governments to adjust their activities to better combat the ultra-processed food industry's influence. Ironically, some of the most effective potential changes copy the industry's playbook. For example, many processed companies often work together and pool their resources to impact government policy making despite being competitors in the marketplace (Moodie et al., 2021). NGOs and public health organizations should try to increase collaboration amongst themselves and with other constituents such as researchers and other influential figures to counteract the industry's influence.

44

Public health efforts should also try to recruit people with a more diverse set of skills to add new elements to their approach, such as digital marketing and personal lived experiences with NCDs (Moodie et al., 2021). While it is impossible to deny the positive effects that this industry has on society, ranging from their contributions to the market to their "philanthropic" activities laced with ulterior motives, we must strive for more widespread recognition of and action against the many negative effects the industry has on public health and the government's ability to effectively serve the interests of their citizens.

6

IMPLICATIONS

By Ipsa Gusain

INTRODUCTION

Processed foods have been around since the early years of civilization to avoid spoilage and ensure resources during periods of drought, but the rapid increase in ultra-processed foods (UPF) in the twenty-first century alone has led to many implications on the global food chain, the healthcare system and overall quality of life. Before discussing these implications, it is important to understand the security, nutritional and consumer benefits that processed foods provide the general population that led to the global UPF empire.

BENEFITS OF ULTRA-PROCESSED FOODS

Food Security & Cost-Effective

With the global number of undernourished people only increasing, especially with COVID-19 worsening the economic state, companies were forced to find ways to produce and distribute more food with fewer resources. Even prior to the pandemic, factors such as population

46

growth, environmental disasters such as floods and climate change amplified food inaccessibility and magnified price concerns all across the world. According to the 2012 Global Hunger Index Report, there are over one billion food insecure people around the world (Weaver et al., 2014). With the focus on invention and innovation of food processing techniques over the past few years, companies and large corporations will soon be able to increase food safety, in addition to the prolong shelf-life and accessibility allowing a large variety of food items to reach populations that were previously struggling with food insecurity (Srour & Touvier, 2020). According to American findings by Lustig, UPFs cost only half as much per calorie as whole foods and thus, the tendency to increase in expenses is lower over time and more sustainable for lower-income families (2017). To many households, the short-term investment is far more beneficial; however, what are the long-term implications?

Nutritional Security

Through the process of fortification—the mechanical or chemical addition of nutrients at a higher percentage than what naturally occurs in the food in question—food engineers were able to ensure people were receiving an adequate amount of vitamin A, vitamin D, vitamin C, thiamin, folate, calcium, magnesium and iron (Weaver et al., 2014). Without processed foods, most of the population would be deficient in at least one of the aforementioned nutrients.

Consumer Benefits

One of the most interesting benefits that came out of food processing technology of the twenty-first century was the ability to create foods that offer health benefits to the consumer. Food engineers have been working on digestion-resistant starches to combat the increasing risk of diabetes and cancer as a result of UPF by slowing digestion while enhancing nutrient bioavailability (Weaver et al., 2014). Other current and future food innovations include altered salt crystal structures to reduce salt intake and greater flavour and nonthermal processing to create stable foods all-year round at cost-effective prices (Weaver et al., 2014). With all these new techniques, and more in the works, processed foods can rebuild a better reputation and therefore, reverse the adverse effects that leave consumers worried about their long-term health.

Globalization

All of these accumulated benefits resulted in UPFs being used globally, a trend that was destined to only increase with time since the initial industrialization of processed foods. Well-packaged, shelf-stable foods were being transported long distances to reach foreign consumers with no second thought given to previous challenges such as storage and food safety concerns (Weaver et al., 2014). UPFs were able to bridge the gap between low and middle income countries, providing safe and stable foods for all.

IMPLICATIONS ON HEALTH

As discussed in past chapters, UPFs have a detrimental effect on health and cognitive capacity. According to Srour & Touvier, many studies have documented strong associations with processed foods and negative health outcomes ranging from obesity to neurodegenerative disorders like dementia (2020). Neurodegenerative diseases (ND) stem from the gut microbiota dysregulation which alter the nervous system through the progressive loss of neurons (Leo & Campos, 2020). This can occur with the constant high in fat diets which affect the microbiota containing short-chain fatty acids that act as neuromodulators involved in neuroplasticity and brain function (Leo & Campos, 2020). The neurologic symptoms include compromised balance, ataxia, speech and breathing (Leo & Campos, 2020). The most prevalent NDs include Alzheimer's disease, Parkinson's disease, Huntington's disease, and spinal muscular atrophy (Leo & Campos, 2020). The past decade has, perhaps not so coincidentally, seen a large increase in ND prevalence; Leo & Campos estimate that over 30 million people have dementia with rates possibly reaching 75 million by the year 2030 (2020).

The cognitive decline associated with chronic UPF diets can also be explained by the hippocampus which is sensitive to tissue damage resulting from the saturated fats (Leo & Campos, 2020). This is because the central nervous system's homeostasis is disrupted; this disruption was not seen in those partaking in a diet of fibre and antioxidants (Leo & Campos, 2020). In early Alzheimer's as well as other neurodegenerative diseases, the hippocampal-dependent memory function is more susceptible to UPFs-related damage as high glycemic foods negatively influence cognition (Harriden et al., 2022).

IMPLICATIONS ON QUALITY OF LIFE

The overall implication of such high fat, simple carb dependent UPF diets is that there is a decline in the quality of life. With life expectancy increasing, a decline in cognition results in unhealthy ageing and the loss of independence. According to Harriden and colleagues, the population of people that are 60 and over will increase to over 20% by the year 2050 with those that are over 80 years old reaching seven billion (2022). There is a rapid increase in life expectancy, but the incidence of chronic illness and cognitive impairment is also increasing just as fast due to nutritional restrictions on diet (Harriden et al., 2022). There is a strong tendency for cognition to weaken with older age which sadly affects the brain's psychological functions such as verbal fluency, judgement, abstract thinking, spatial orientation etc. (Harriden et al., 2022). While there is an intermediary stage between having normal cognition and dementia labelled Mild Cognitive Impairment (MCI) where not everyone in this category progresses to dementia, food-dependent cognitive decline is also considered a preventable disease by adopting better dietary habits (Harriden et al., 2022). The scary truth of dementia is that there are about 50 million people around the world with this impairment with about 60-70% dealing with Alzheimer's disease (Harriden et al., 2022). This has so many implications on the individuals themself including physical and mental health but also on their families and the economy which in turn impacts global healthcare systems. As there are no effective pharmaceutical treatments for dementia, it is important people find ways to mitigate this health risk by adopting better eating habits full of adequate nutrition to live a happier and more fulfilling life for much longer.

Unfortunately, with the rise of accessible UPFs worldwide, children are being exposed to processed foods at a much younger age than ever before. The first two years of life are key for young children to learn and build long-term healthy dietary behaviours but the reality for many low-income families and communities is that they do not have an option to feed their children fresh food (Relvas et al., 2017).

A study done by Relvas and colleagues monitored the feeding practices of children aged six to twelve months that live in São Paulo, Brazil. They recorded frequent consumption of UPFs amongst the young infants despite NOVA recommendations that children under 2 not consume UPFs (Relvas et al., 2017). NOVA is a food classification system based on how much processing a food went through; it has four categories ranging from unprocessed to UPFs and drinks (Gibney et al., 2017; Relvas et al., 2017). NOVA has been integrated into numerous major international diet and health reports with many national governments following suit (Gibney et al., 2017). UPF consumption is an early determinant of chronic disease so to conclude, children are continually being exposed to a diet that puts them at risk of chronic disorders at a much younger age. Partaking in UPF diets from a young age may result in more accumulation of the harmful effects of processed foods and may have life-long risks that develop much sooner than currently observed.

IMPLICATION ON HEALTHCARE

The cost of the global health epidemic that is a direct result of the high UPF consumption is still a relatively new issue in the healthcare system. This is because only within the past century has large-scale food processing been industrialized and thus, professionals are still researching and learning the long-term effects of processed foods (Gibney et al., 2017).

What is becoming increasingly obvious to professionals and consumers alike, is the large sum of healthcare expenses that accompany the chronic illnesses and disorders that form from poor dietary habits. As mentioned earlier, the short-term investment in cheap UPFs is seemingly more sustainable for lower-income families. However, in the long-term, these families will have to spend thousands of dollars on insurance premiums along with the reduction in income from reduced work years from disability and increase in life expectancy making it the long-term, less sustainable option (Lustig, 2017). The implication of UPFs on health care can be demonstrated by the gross domestic product (GDP) growth over the years; from 1965 to 2014, the GDP increased from 2% to a staggering 17.9% (Lustig, 2017). Healthcare costs about $3.2 trillion annually from which 75% is spent on metabolic syndrome diseases which are preventable with a more nutritious, low in fats food and better dietary habits (Lustig, 2017). Alongside the expenses, healthcare is not accessible to all and can be difficult to find in low-income countries and for low-income populations.

IMPLICATION ON INEQUALITY

What is currently being seen in wealthier countries such as Canada and the U.S. is a shift towards the desirable whole foods, at least for higher-income families and populations. Those that can afford to buy locally grown foods do so as there has been a greater advocacy towards the health benefits and the rise of the Support Small businesses movement. Unfortunately, not everyone can afford to shop locally so the health gap between healthy, rich citizens and less healthy, lower-income citizens is only increasing. For example, students not living with their family often buy more processed foods than produce because they are more shelf-stable and less susceptible to spoilage. Many times, these students don't have a choice in the matter due to time and financial limitations; most are well aware of the harms of these UPF diets as schools all over the globe are becoming more proactive about speaking on the dangers of processed foods.

The past decade has definitely seen the rise of instant ramen for this reason. Ramen has become the new superfood that alleviates hunger of low-wage workers and students in rich and poor countries alike (Goodman, 2016). In fact, according to Goodman, ramen consumption has surpassed McDonald's consumption as well as other processed foods as a whole where even many humanitarian food aid packages have begun to include it as it is incredibly filling for a long period of time (2016). While having a resource like ramen around the house for a rainy day is always reassuring, there is no doubt that UPFs like these instant noodles, are not helpful in reducing starvation in lower-income countries. In fact, Goodman argues that it only masks starvation in those that need more food security (2016). Low-income neighbourhoods can globally be defined as having an absence of grocery stores that carry fresh foods with a higher percentage of fast food outlets and convenient stores carrying UPFs (Goodman, 2016). By having virtually no options to shop healthier, these populations have had their choice to live a long and healthy life taken away.

IMPLICATION ON GLOBAL FOOD SYSTEM

The implication of UPFs on the global food system has been tremendous over the last century. The globalisation of the food chain has made large transnational food corporations' major players in the food industry worldwide with some of the biggest food corporations being Coca-cola, Pepsico, Kraft, Nestle etc. (Gibney et al., 2017; Lustig, 2017). These corporations know that regardless of the growing concern with flavourful, convenient, and affordable foods over the optimal diet of whole foods, UPFs have been dominating due to the economic state of the world and the cultural shift towards homogeneity. Ritzer, author of The McDonaldization of Society, describes the societal shift towards decision-making centred around the factors of predictability, control, calculability, and efficiency (Goodman, 2016). The world is quickly adopting the need for homogenization and standardisation as seen in institutions such as education and healthcare (Goodman, 2016). When it comes to the global food chain, people want dependable service that does not stray from the normal. Realistically, this can only be achieved with overly processed foods and thus, big corporations will continue to favour UPFs over nutritional whole foods. For there to be an everlasting change in food consumption and global dietary habits, realistic policies and programs to improve public health need to be established and implemented with the help of these large multinational companies partnering up with small and medium-sized enterprises (Gibney et al., 2017).

IMPLICATION ON FUTURE POLICIES & SOLUTIONS

One of the solution-based implications of food-related cognitive decline is a better labelling system or Front of Package (FoP) labelling. Like cigarette packages, UPFs have a required safety warning on each product (Taillie et al., 2020). These labels serve the purpose of helping the consumer make informed decisions about the products they are buying and consuming. Even with a general knowledge of how unhealthy processed foods are to long-term health, the skyrocketing demand of UPFs have proven that the general public does look past scientific research and statistics for short-term comfort. These new FoP labels will help discourage the public from excess consumption by eliciting a negative effect and a greater perception of risk (Taillie et al., 2020). The government is calling for a major behavioural change from the public. According to Taillie and colleagues, these warning labels will only be effective if they grab the attention of the public and are accurately understood by the vast majority (2020). A lot of research has gone into greeting eye catching and informative labels highlighting the global health crisis that processed foods have created in the past century alone.

While one of the biggest benefits that have come out of the easy accessibility and long shelf-life of UPFs is worldwide access to a variety of foods, recent studies demonstrating the health impairments with malnourishment from UPFs show an increase in universal obesity, sedentary lifestyle and chronic degenerative noncommunicable diseases–CVD, diabetes, chronic respiratory diseases etc.– rates (Weaver et al., 2014). The implication of creating processed, accessible foods, is that even in areas of low food security, having enough of these types of food is no longer considered a good diet (Weaver et al., 2014). Therefore, food and nutrition scientists have been researching ways to provide nutritious and healthy processed foods, much of which includes the innovative food processing technologies discussed earlier in the chapter. The answer is multifaceted as the solution is not as simple as making fresh, local foods accessible to larger populations. Just like how the current processed food in its current state can't meet all nutritional requirements, fresh and local foods cannot meet all requirements either (Weaver et al., 2014). Therefore, health professionals and food engineers must continually look towards improving processing technologies by reducing saturated fats, sodium and added sugars while minimizing nutrient losses.

IMPLICATIONS OF COVID

With the dominating COVID-19 pandemic imposing social-distancing policies and disrupting the economy, the world saw an even bigger spike in risks for chronic diseases such as type 2 diabetes, hypertension and CVD due to startling rates of food insecurity (Leddy et al., 2020). The health disparities as a consequence of unequal access to resources like food, employment, healthcare etc. within communities of colour and lower-income households caused food insecurity to skyrocket and poor health outcomes to only increase (Leddy et al., 2020). COVID-19 brought structural changes that most countries were not prepared for. The unemployment rate in the U.S. alone, increased from 3.5% to over 14% with those that weren't laid off having reduced hours and subsequently, income (Leddy et al., 2020).

The pandemic and its accompanying policies resulted in the overnight shutdown of nonessential institutions including restaurants, schools and many worksites across the globe. COVID revealed the fragile food system; the closures led to a rapid, and unsustainable, demand for home food and food inequalities in grocery stores (Leddy et al., 2020). Many farmers were also forced to throw away an unfortunate amount of food that should have gone to restaurants, fast food restaurants and other institutions of consumption outside of the house (Leddy et al., 2020). Overnight, grocery stores, who are the biggest contributors to food banks, had food shortages with people purchasing and panic stockpiling (Leddy et al., 2020). No longer able to provide food to food banks, food banks began purchasing more food leading to inflation of foods with millions of U.S. citizens depending on these banks receiving even less than prior (Leddy et al., 2020).

The health gap has implications on medical adherence as people with low adherence may be doing so to save money. This behaviour has always existed in lower-income communities but due to the pandemic, rates of poor medical compliance have increased (Leddy et al., 2020). Poor medical adherence is, similar to UPFs, not cost effective in the long-run with many diseases and conditions only worsening with non-compliance and treatment bills costing more than the initial deposit. People with low medical adherence were often those relying on cheaper, more energy-dense foods due to food insecurity (Leddy et al., 2020). Further, COVID has stripped millions of people worldwide of their health and overwhelmed hospital units; many people were left with no access to doctors or medical help unless deemed a life-threatening emergency, even in higher-income countries. Therefore, more people than ever could not get any medical help for developing diseases that could be traced down to high in UPF diets.

CONCLUSION

The implications of the worsening health of high ultra-processed food consumers have matched the benefits of having accessible foods with a long shelf-life. Highly processed foods have health implications on the gut microbiota and the hippocampus, detrimental effects on the quality of life at a time where life expectancy is at its highest, led to healthcare becoming more inaccessible than ever and caused a greater division in low versus high-income populations. Food scientists and engineers must work towards not eliminating processed foods, but creating new technologies that will undo the damage of having such innutritious, energy-dense, and high-in-fat foods.

7

THE ROLE OF HEALTH AND ENVIRONMENTAL ETHICS IN THE PROCESSED FOOD INDUSTRY

By Lydia C. Rehman

INTRODUCTION

Chronic diseases such as diabetes, cardiovascular disease, cancer are the leading cause of death around the world (Schermel et al., 2014). Poor nutrition is the result of many chronic diseases, including high/low blood sugar, and obesity (Schermel et al., 2014) . Poor nutrition and chronic diseases directly contribute to an individual's quality of life, and may reduce their overall life expectancy (Schermel et al., 2014). The annual cost of obesity in Canada is nearing 5 billion dollars (Schermel et al., 2014). Overall, there are major health, and socio-economic impacts caused by the production of ultra-processed foods. Throughout time there has been a shift in the way food is preserved/processed, and this has now become an industrial process that has lasting impacts on the safety, quality and nourishment that it provides its consumers (Monteiro et al., 2015). Further, ultra-processed foods are not balanced in their nutritional benefits, due to the ways in which they are produced, the industrialization of food processing has altered the natural ways foods are derived (Monteiro et al., 2015) . Many ultra-processed foods are marketed to be more visually pleasing and they are consumed in high quantities, and often displace less processed foods that use fresh ingredients (Monteiro et al., 2015).

Producing ultra-processed foods has a profound impact on culture, economy, environment, public health, and cognitive brain functioning (Monteiro et al., 2015). For example: poor health outcomes are directly associated with high consumption of sugary drinks, and processed foods, yet they remain highly available to populations in both food, and non-food retail businesses (Mezzacca, 2020). These ultra-processed foods make it easy for the everyday consumer as most products are ready for consumption with minimal preparation (Monteiro et al., 2015). Overall, given the the burden of chronic disease that is attributed to poor nutritional intake; this paper aims to emphasise the various ethical considerations within the industrialization of the ultra-processed food industry. Likewise governance and public health policy is critical to addressing current disparities impacting population health.

MORAL RESPONSIBILITY & LONG-TERM HEALTH

It is necessary for all living beings to consume food, however given this basic need involves risk with food safety, and can take years to impact an individual's health (Potthast & Meisch, 2012). For instance; one of the main reasons for obesity is correlated with high caloric intake, and storage of excessive food in the body that turns to fat. An extended period of excessive eating leads to insulin resistance, and ultimately predisposes individuals to contract type 2 diabetes mellitus, or cardiovascular disease (Potthast & Meisch, 2012).

As this phenomenon may occur later in life with the onset of chronic disease, there is a strong correlation between safety and health related issues. There must be an understanding of what these dynamics are and the role of all stakeholders in the ultra-processed food industry (Potthast & Meisch, 2012). Those responsible for producing food for consumers are required to alert if there is harm that could become potentially hazardous to human health. For example: the U.S Food and Drug Administration has legal obligations of its producers to ensure that any risks such as: best before dates, levels of salmonella are shared with its distributors, and consumers. However, the regulatory system of the food industry fails to recognize that food related damages to health is not a short-term phenomenon.

These specific health damages (eg. chronic disease, obesity) can occur over an extensive period of time leading to poor health status and lessened quality of life. From a financial standpoint, obesity in the United States is a huge revenue to their privatised healthcare system as medical costs that are related to obesity are around 150 billion annually (Potthast & Meisch, 2012). Given the macro-scale of processed food production, knowing that there are tremendous risk factors that can impact long-term health. Establishing who is morally responsible and contributes to the continued health damage is important for establishing long-term sustainable solutions (Potthast & Meisch, 2012). Given the immense revenue that obesity causes, and the poor health status that is attributed to poor nutrition, the intent of the corporations that produce ultra-processed foods is highly questionable (Potthast & Meisch, 2012). Processed food industries should ensure that all decision making is done through a lens of ethics, and with the intent of protecting, and preventing poor health status for its consumers (Potthast & Meisch, 2012). Quality assurance, responsibility and regulations should be put in place to protect human health. Further companies should aim to reduce their production of high sugar/sodium processed foods.

Those in the field of research and public health have explored the social and economic costs of consuming ultra-processed foods (Rossi et al., 2017). There needs to be greater advocacy, and health promotion for increasing accessibility to local/fresh produce (Rossi et al., 2017). Reducing ultra-processed foods and introducing increased vegetables is important for improving health outcomes. Based on the existing literature, corporations do not take full responsibility for causing health related issues such as obesity; they are merely just associated with it (Rossi et al., 2017). Further the issues related to ultra-processed foods that affect society are systemic and are caused by various actors that perpetuate harms through action, each actor should be held at some standard of responsibility. One research initiative looks at community supported agriculture (CSA), these programs are designed to transform the relationships between people and food. Additionally, these programs offer seasonal subscriptions to farm products and in exchange the consumers get locally sourced fresh goods. Overall, these programs aim to change the means of production, distribution and alter the consumers eating and purchasing behaviours (Rossi et al., 2017). This program allowed the farmers to be in control of providing locally grown produce, and the results of the study found that the participants who used the CSA programs were more likely to continue purchasing food directly from farmers (Rossi et al., 2017). CSA's is one of many evidence-based solutions that help drive consumer behaviour changes related to food. CSA's should be further researched to better understand its impacts and benefits for society and the environment, given the multitude of disease, environmental impacts caused by production of ultra-processed foods (Rossi et al., 2017).

In North America, the food industry is controlled by large organisations that are unconcerned about the environmental, economic and societal impacts of processed foods (Potthast& Meisch, 2012). These companies are only concerned with the profits they generate, and the unfortunate reality is the high production of sugary foods, the extensive use of artificial ingredients and additives, and genetic engineering to produce on a macro-scale. These corporations should be liable and regulated to provide consumers, and distributors with all the necessary information to make informed healthy choices. Corporations should be operated to ensure that any and all health hazards that their consumers face should be addressed, including potential long-term health impacts caused by ultra-processed foods (Potthast & Meisch, 2012). The food production system has the ability to ensure they educate customers on healthy eating patterns that drive nutritional value (Potthast & Meisch, 2012).

ENVIRONMENTAL IMPACTS OF FOOD PRODUCTION

While food processing works on various levels, ultimately processing is just one component that makes up an entire food system (Klosse, 2019). Ultra-processed products are a major cause behind environmental degradation, pollution, loss of energy, and contribute to a high degree of waste (Monteiro et al., 2015). These systems work to produce, store, sell, transport, consume, and waste food related goods (Klosse, 2019). These processes are directly correlated with health, well-being, environmental impacts and overall nutritional value (Klosse, 2019). Global populations are faced with being under/over nourished due to the impacts of the way foods are produced and consumed (Klosse, 2019). While corporations are highly focused on maintaining their consumers through providing sugary, and high sodium products. Another ethical concern within the processing of food needs to begin at the farm level (Klosse, 2019).

Due to the demands of increasing global populations, farmers have been noted to use industrial chemicals that are directly associated with climatic impacts such as: global warming, loss of biodiversity (Klosse, 2019; Rowan, 2020). Current crops have been genetically engineered and patented by large biotechnology companies (Klosse, 2019). There are high quantities of agro chemicals spraying of crops to minimise (pests, and weeds). While this seems innovative, the result is a high degree of land degradation, groundwater contamination, and the extensive use of fertilisers that require non-renewable energy (Klosse, 2019). Further the use of glyphosate which is a main chemical in insect/herbicides have been tremendously hazardous to human health; as residues from crops move from the farm onto the consumer (Klosse, 2019). The hazards of glyphosate are carcinogenic to humans, and major reformation of the food industry needs to occur, as this is a huge ethical concern impacting human health (Klosse, 2019). As previously mentioned, food waste is extensively high, inadequate practices within farming are critically hazarding the environment (Klosse, 2019).It is evident that standardised agricultural practices that use chemicals, have shown to be a major hazard to human health (Abushal, 2021). There are cases of cancer, miscarriage, and birth defects. Many of the chemicals that are used lack the necessary health information, and legislation is non-existent to regulate use of such chemicals and practices (Abushal, 2021).

With high levels of population growth, there is an increasing demand for food, and with demand comes increased food prices, food insecurity, climate change and undernourishment. While biotechnological innovation has the capability of improving food security, by supplying more communities with greater quantities of food, it is not without ethical concern (Abushal, 2021). Small farmers are directly impacted by monocultures of genetically modified crops, and can genetically pollute organically grown crops (Abushal, 2021). The companies in control of regulating crop production with genetically modified crops are not efficiently regulating the use of agro-chemicals which ultimately results in the loss of biodiversity. For example: some Mayan communities in Mexico have been severely impacted by the presence of GM crops, as their organic honey production produces less. Further, the bees used to produce honey are diminishing in population size, as plants that the bees pollinate are dying, and becoming toxic to its micro-consumers (Abushal, 2021). The interesting point of all is that these companies have the potential to feed communities that have faced high levels of food insecurity, yet they are not required to label their goods as genetically modified (Abushal, 2021). Further these corporations have a major obligation to ensure that their consumers can trust in the safety of their products.

Corporations such as 'Monsanto' amongst other large organisations control all commercially available seeds (White, 2020). Corporations have full control of the genetic engineering of the seeds, they control the carcinogenic materials that are used on them, in which making something that should grow and flourish with the earth, toxic to the species that rely on these foods, as was the case for the Mayan bees (White, 2020). Their large monocultures are just another genetically engineered display of colonialism and capitalism, and are responsible for the loss of biodiversity, at the expense of community health for profit (White, 2020). These companies are constantly disempowering farmers, by suing them when farmers try to improve the quality and variety of seeds (White, 2020). Ultimately corporations are responsible for food insecurity, and their attempts disempower traditional knowledge that Indigenous communities that have protected and passed down through generations for thousands of years (White, 2020). Given the unethical management of food production at the micro-level; solutions need to ensure that Indigenous worldviews, ceremony are included in the process of reclaiming traditional food systems, as well as creating sustainable solutions that are good for human and environmental health.

61

FOOD SYSTEM TRANSFORMATION, COLONISATION, AND RECLAIMING TRADITIONAL FOOD SYSTEMS

Indigenous food systems consist of traditional foods, seeds, practices that are sustainable, and carbon reducing in nature (White, 2020; Settee and Shukla, 2020). Old seed varieties have increased proteins, antioxidants, amino acids, yet climate change and issues associated with food production will continue to be problematic for the next hundreds of years because of the role of industrialization, capitalism (White, 2020). Further, as a consequent result of colonialism, and the systemic barriers: such as infrastructure, land displacement, the burden of disease have impacted Indigenous populations from having the ability to be able to sovereignly cultivate traditional crops such as the sunflower, corn, squash. It is argued that to strengthen ties with Indigenous populations; communities, the land, First Nations, Metis and Inuit populations should be able to utilise their traditional foods, and knowledge to have self-determination and food sovereignty in a place where colonialism has had a profound impact on Indigenous population health (Settee and Shukla, 2020). Given the health benefits of traditional seeds for production and processing, with the incidence of illnesses such as cardiovascular disease, diabetes, hypertension, food insecurity, revitalization is important to re-establish balance for healthy health, mind, and body (Settee and Shukla, 2020). Indigenous populations have the knowledge and resilience to be able to live through a realm of wellness rather than illness. There needs to be greater emphasis on being able to care for the land without the systematic colonial barriers, which will only allow Indigenous populations to reclaim sovereignty and self-determination. Overall, Indigenous foodways, and knowledge systems, should be appreciated and celebrated for the profound impact that they provide in terms of health, wellness, environment and the cultural diversity they share (Settee and Shukla, 2020).

One research study explores Canadians' perceptions towards diet (Schermel et al., 2014). The results of the survey found finding healthy processed foods was of major difficulty as healthy foods come at higher costs. Likewise, finding healthy processed foods is difficult, although the consumer would find interest in purchasing these products. (Schermel et al., 2014). Consumers have a right to be educated on being able to identify healthy foods, solutions may include creating subsidies for fresh produce (Shermel et al.,2014). In addition to the barriers of accessing healthy foods, accessing healthy "processed" foods was also difficult as foods contain too much salt, sugar or lack nutritional value. Further a high percentage of respondents felt as if their diet was enriching and did not need to be changed (Shermel et al.,2014). Likewise, It is worth noting that over 1 billion people globally are considered to be undernourished, lacking the necessary vitamins within their foods and diet. It is globally understood that greater increases in plant-based diets that are not ultra-processed is significant to reducing sugar and sodium intake (Klosse, 2019). Further creating patterns within individual diets that are more sustainable, would reduce chronic disease, prevent undernourishment, and reduce greenhouse gas emission is critical to human and environmental health (Klosse, 2019). Targeting food insecurity moves far beyond looking at it from the standpoint of morality, or policy, rather having a right to food is a human right (Klosse, 2019).

CONCLUSION

Overall, corporations, policy-makers, and public health organisations have an obligation to ensure that the food industry is regulated to reduce the onset of chronic diseases, and overall reduce the obesity epidemic that currently exists. Societies are either under or over nourished, and either way are not meeting the daily recommended servings (Schermel et al., 2014). Given the complexity of the issues discussed, all stakeholders within food processing need to work on all levels to address the growing socio-economic, health, and environmental factors that the industry holds. Further education should be provided to influence healthy food patterns, to promote healthy sustainable living. Legislation should ensure that every person is able to purchase nutritious food accessibility without the burden of high costs (Schermel et al., 2014).

8

PRACTICAL ACTIONS TO ASSISTING COGNITIVE DECLINE AND HEALTHY HABITS

By Uzair Tazeem

INTRODUCTION

Advancements in technology in the last century have drastically changed the landscape of healthcare resulting in improved life expectancy and overall health outcomes. Although people are living longer than before, our society now faces the economic and social burden of age-related illnesses which must be addressed (Wang et al., 2019). Age remains the number one risk factor for cognitive decline, which refers to progressive and gradual changes in brain function and structure that disrupt cognitive function and daily activities (Shatenstein & Barberger-Gateau, 2015). Based on the severity, there are three degrees of cognitive decline: normal age-associated cognitive decline, mild cognitive impairment (MCI) and dementia. Those who suffer from age-associated cognitive decline may have trouble remembering names or locations, whereas MCI patients would experience more severe short-term memory challenges but remain functionally independent. Dementia in comparison, refers to more severe memory impairment and loss of intellectual function, which interferes with the person's daily activities and may require additional assistance (Barrett et al., 2021). It is important to understand that although age is a risk factor for dementia, dementia is not an inevitable outcome of aging. This is because synergistic interactions of key risk factors can accelerate cognitive decline and lead to the development of neurodegenerative disorders such as MCI, Parkinson's disease, Alzheimer's disease (AD), and other forms of dementia (Lin et al., 2017). On the other hand, this suggests that management of risk factors can potentially slow down cognitive decline and more severe forms of cognitive impairment. AD is the most common form of dementia, making up nearly 60-80% of all cases. For those aged 65-74, the risk of Alzheimer's and related disease (ADRD) is almost 3%; however, by the age of 85, this risk increases to a staggering 32%. This is very worrying because not only is ADRD one of the costliest diseases in America, but it is also among the 6th leading causes of death for which there is no cure (Barrett et al., 2021). Although there are no effective treatment options available, there is a growing body of evidence to suggest that modification of lifestyle risk factors can build, preserve, and even enhance cognitive function later in life. It is important to understand that before the onset of dementia and its symptoms, there is decades-long accumulation of pathological changes that disrupt brain function (Shatenstein & Barberger-Gateau, 2015). This review aims to provide insight into the risk factors associated with cognitive decline and provides a practical set of actions that can be taken to reduce the risk of dementia and other related illnesses.

From a clinical standpoint, there is great variability in the rate of cognitive decline and its symptoms. (Shatenstein & Barberger-Gateau, 2015) presents the cognitive reserve hypothesis to explain this variability in individuals affected by the same neurodegenerative changes. The hypothesis suggests that higher levels of cognitive reserve related to a healthier lifestyle acts as a protective mechanism that delays the rate and symptoms of cognitive decline. The cognitive reserve hypothesis is in line with many studies that link a healthier lifestyle with favorable cognitive health outcomes. Cognitive health is extremely important for the overall health and well-being of the elderly population; therefore, it is imperative to understand better the risk factors which accelerate cognitive decline. Many risk factors associated with cognitive decline can be categorized into two main categories: non-modifiable and modifiable risk factors.

NON-MODIFIABLE RISK FACTORS

Non-modifiable risk factors cannot be altered by medical intervention or individual effort (Baumgart et al., 2015); the two most common are age and genome. As discussed earlier, age is directly related to cognitive impairment as the buildup of neurodegenerative pathologies overtime alters the brain's function and structure. Therefore, to an extent, age-related cognitive decline is inevitable. In addition, genetic make-up can make some individuals more susceptible to cognitive decline than others. Alzheimer's disease follows an autosomal inheritance pattern wherein mutation in the presenilin or amyloid precursor protein results in the disease. Furthermore, many studies have suggested that mutation in the APOE-4 gene, which codes for apolipoprotein, greatly increases the risk of ADRD (Small, 2016). Although age and genetic make-up are non-modifiable risk factors, it is important to have a thorough understanding as these factors can be used for early detection of cognitive decline.

MODIFIABLE RISK FACTORS

Modifiable factors, on the other hand, can be changed at the level of individual to reduce the risk of cognitive decline. Due to the fact that illnesses related to cognitive decline have no effective treatment options or cure available, reduction of risk factors is the best form of medication. As with any disease, prevention is always better than treatment. The main modifiable risk factors are categorized into four categories: collective societal, individual psychosocial, lifestyle, and cardiometabolic factors (Shatenstein & Barberger-Gateau, 2015).

SOCIETAL FACTORS

The collective societal factors refer to factors which characterize an individual's environment at the macro-level determining one's access to education, healthcare, and health amenities. These factors can influence an individual's lifestyle choices, which may accelerate the rate of cognitive decline (Shatenstein & Barberger-Gateau, 2015). A study in Quebec by (Gauvin et al., 2012) found that environments play an important role in fostering a healthy and more social lifestyle for seniors. Researchers found that greater accessibility and availability to services and amenities was associated with more frequent walking among all adults. On the other hand, disadvantaged neighbourhoods that lack resources and facilities lead to physical inactivity and a sedentary lifestyle (Teychenne et al., 2012).

PSYCHOSOCIAL FACTORS

Psychosocial factors include an individual's socioeconomic status, level of education, social network, etc. These factors define one's characteristics, behaviors and circumstances in life and therefore directly impacts cognitive reserve (Shatenstein & Barberger-Gateau, 2015). Many population-based studies have shown a strong relationship between socioeconomic status and cognitive impairments that lead to dementia. It is seen that individuals who come from a low socioeconomic background often suffer from preventable health conditions such as hypertension, diabetes, cardiovascular disease, etc which are all risk factors for cognitive decline (Shatenstein & Barberger-Gateau, 2015). A study by (Wen et al., 2014) showed that low-income households with low levels of education have a high sugar/fat/protein diet. Not only is it more difficult for these families to afford healthier foods, but they also lack knowledge on appropriate nutrition for a healthy lifestyle. Education is a very important determinant of overall health as higher levels of education provide the opportunity for higher income and, therefore accessibility to healthcare, higher quality of diet and lifestyle (Shatenstein & Barberger-Gateau, 2015). A meta-analysis in the UK between 1994 and 2004 found that women who left school at an early age were at a greater risk of developing dementia than those who pursued higher education levels (Russ et al., 2013).

LIFESTYLE FACTORS

The choices we make and the activities we partake in daily make up the lifestyle factors that are directly associated with a cognitive reserve and the trajectory of cognitive decline later in life.

Unlike societal and psychosocial factors, lifestyle factors are under our immediate control and can actively be worked on to improve cognitive health. These lifestyle factors include nutrition, physical exercise, cognitively stimulating activities and stress management (Shatenstein & Barberger-Gateau, 2015).

NUTRITION

From the very beginning, adequate nutrition is extremely important for normal fetal brain development. Nutritional deficiency, particularly in long-chain omega-3 fatty acids, vitamin B, zinc, and iodine, can disrupt the development process and result in irreversible changes in brain structure and function. Fetal malnutrition can result in lower child intelligence and prevent higher attainment of education that could lead to an unhealthy lifestyle when older thereby accelerating cognitive decline. Nutrition is a very significant determinant of cognitive health at old age, and for this reason, many diet patterns have been thoroughly studied to better understand its relationship with cognitive health (Shatenstein & Barberger-Gateau, 2015).

The Mediterranean diet has garnered great attention in recent decades as evidence shows that it reduces the risk of many metabolic disorders such as diabetes, hypertension, obesity, etc which are all risk factors that accelerate cognitive decline (Small, 2016). The Mediterranean diet consists of high consumption of fish, olive oil, nuts, fruits, vegetables, and whole grains while minimizing fast/fried foods, pastries, butter, and cheese (Shatenstein & Barberger-Gateau, 2015)). Olive oil,fruits and vegetables are rich in antioxidant compounds which help to remove free-radical compounds and reduce levels of oxidative stress in the body (González-Gross et al., 2001). A study by (Pitchumoni & Doraiswamy, 1998) highlighted that individuals with Alzheimer's are often present with high oxidative stress levels caused by antioxidant imbalance. Additionally, omega-3 fatty acids from fish and nuts have been shown to have an anti-inflammatory effect and monounsaturated/ polyunsaturated fatty acids help maintain the structural integrity of neurons (González-Gross et al., 2001). Furthermore, evidence suggests that diets rich in polyunsaturated fatty acids not only delayed the development of cognitive impairment but also improved participants' cognitive performance (Shakersain et al., 2015). In comparison, a western diet is characterized by intake of red/

processed meat, saturated/trans fats, refined grains, and high sugar content which is associated with greater rates of cognitive decline. The reason for this is that a high-calorie diet that is rich in saturated fats, and sugar accelerates the development of metabolic disorders such as diabetes, obesity, and hypertension, all of which are risk factors for cognitive decline. Furthermore, a western diet includes heat-processed foods containing high levels of advanced glycation end products (AGEs) (Shakersain et al., 2015). (Hsu & Kanoski, 2014) has demonstrated that AGEs are associated with the amyloid (Ab) peptide, which degrades the blood-brain barrier and causes accumulation of amyloid (Ab) in the hippocampus of the brain. This is believed to be the primary mechanism of action which accelerates the rate of cognitive decline amongst individuals who adhere to the western diet.

PHYSICAL ACTIVITY

Similar to nutrition, physical activity is a key determinant of cognitive health as many risk factors associated with dementia, such as diabetes, hypertension, and obesity, are linked to physical inactivity (Shatenstein & Barberger-Gateau, 2015). This is further supported by a study by (Norton et al., 2014) which found that the vast majority of AD cases in the US, Europe and the UK are the result of physical inactivity. Exercise is believed to elevate levels of brain-derived neurotrophic factors (BDNF) which has a multitude of effects on the brain. BDNF repairs damaged neurons, stimulates neurogenesis to create new neurons, increases blood flow through angiogenesis, and stimulates branching of dendrites via synaptogenesis, thereby improving cognitive function (Small, 2016). Angiogenesis, neurogenesis, and synaptogenesis help to counteract age-related changes to overall cognitive health and lower the risk of cognitive impairment.

COGNITIVELY STIMULATING ACTIVITY

To preserve cognitive function at old age, it is important to participate in cognitively stimulating activities, which many studies suggest, help to strengthen neural connections and delay the onset of dementia (Small, 2016). Cognitively stimulating activities can include playing cards or board games, reading, doing puzzles, and attending social organizations and this is associated with reduced risk of cognitive decline (Shatenstein & Barberger-Gateau, 2015). Study by (Dartigues et al., 2013) explored the relationship between cognitive health and individuals who play board games and found that over 20 years, the prevalence of dementia was lower in board game players. This suggest that participating in cognitively stimulating activities helps maintain cognitive health and function.

STRESS

In the last few decades, great importance has been placed on stress management and mental health; this is a great step in the right direction. In the context of cognitive health, (Small, 2016) shows that stress hormones such as cortisol increase the inflammatory response which damages the brain and therefore increases the risk of cognitive decline. Fortunately, however, these changes to the brain can be reversed once cortisol levels return to normal. Relaxation exercises such as meditation, deep breathing and yoga are excellent ways to combat stress and memory abilities. A study by (Eyre et al., 2016) found that yoga not only helped alleviate symptoms of depression but also improved visuospatial and verbal memory, which is related to overall cognitive health. Another effective strategy for reducing stress and thereby improving cognitive function is to get an adequate amount of sleep each night. During sleep, a very important process of memory consolidation takes place, improving our ability to recall and remember information. Moreover, a powerful tool for combatting stress is maintaining a strong social network of friends and family that one can talk to about problems one may face. Several studies have shown that individuals with strong social networks have a longer life expectancy (Small, 2016).

CARDIOMETABOLIC FACTORS

Lifestyle factors, specifically nutrition and physical activity, are important determinants of cardiometabolic conditions, which in turn impact cognitive health. Cardiometabolic conditions refer to a cluster of preventable metabolic disorders that includes type II diabetes, hypercholesterolemia, hypertension, obesity, and others (Shatenstein & Barberger-Gateau, 2015). Many studies have related diabetes, which is characterized by elevated blood sugar levels, to dementia (Baumgart et al., 2015). A meta-analysis study by (Cooper et al., 2015) found that the cognitive health of individuals with mild cognitive impairment and type II diabetes was more likely to deteriorate and lead to dementia than those without diabetes. It is believed that cognitive impairment results from abnormalities related to type II diabetes, such as neuropathy, retinopathy, hypertension, obesity, and others (Nooyens et al., 2010). By causing small-vessel disease and white matter lesions in the brain, hypertension is believed to also cause cognitive impairment and accelerate the rate of cognitive decline . Moreover, hypercholesterolaemia is another risk factor related to cognitive dysfunction as cholesterol-rich diets are shown to increase the deposition of b-amyloid plaques in the brain, impacting function (González-Gross et al., 2001). Obesity has a whole range of effects throughout the body which impact an individual's overall cognitive health. An increase in adipose tissue in the body is associated with reducing the volume of various brain parts. In a study with 1400 Japanese participants by (Taki et al., 2008), researchers found that increase in body mass index (BMI) was associated with reduced volume of the temporal, occipital and frontal lobes and parts of the cerebellum. Many of these brain regions play an essential role in memory and memory consolidation therefore, it is believed that volume reduction is one of the mechanisms of cognitive dysfunction. Additionally, morbidly obese individuals in comparison to non-obese individuals, are found to have elevated levels of amyloid protein in the blood which is also believed to accelerate the rate of cognitive decline(Nguyen et al., 2014). The blood-brain barrier (BBB) plays an important protective role in the central nervous system by regulating the entry of compounds and foreign bodies. Obesity in mid-life is related to the reduced structural integrity of the BBB, thereby increasing the risk of Alzheimer's and other forms of dementia (Nguyen et al., 2014). Although cardiometabolic disorders increase the risk of cognitive impairment, it can be controlled by incorporating healthy habits into our life.

PRACTICAL ACTIONS TO REDUCE COGNITIVE DECLINE

Although cognitive decline is an inevitable and natural outcome of aging, some steps can be taken to slow, preserve and even enhance cognitive function in old age. This literary review has provided an overview of the risk factors associated with cognitive function. Note that many of the risk factors discussed are interrelated with one another, and for this reason, a multivariate approach is required to reduce the risks of cognitive impairment effectively (Baumgart et al., 2015). For example, nutrition and physical activity are both important determinants of the cardiometabolic health of individuals. Therefore, focusing on a single change in lifestyle or health factor may be inadequate in reducing the risk factors associated with cognitive health. For optimal results, we believe that effort must be made both at the individual and the government level. With the global aging population increasing, age-related cognitive disorders pose a great challenge to our economy and the healthcare and welfare system (Wang et al., 2019).

LEVEL OF GOVERNMENT

Government intervention can help improve the cognitive health of populations by first investing in educational programs to raise awareness of cognitive decline and healthy lifestyle habits to reduce risk. Additionally, efforts must be made to improve accessibility to higher education, healthcare and other amenities that directly influence lifestyle factors. For example, it would be difficult for individuals that come from a low socioeconomic background to pursue high education if there is no tuition aid. By providing loan options, scholarships, and financial aid, governments can help improve and encourage education accessibility. Additionally, strategies must be employed to help alleviate the burden felt by those from a lower socioeconomic background, as these individuals often lack the resources and/or education needed to make healthy lifestyle choices to improve cognitive health. Lastly, we recommend constructing an environment that fosters a healthy lifestyle for seniors. For instance, a well-planned and serviced environment would encourage social participation from community members and physical activity that can reduce cognitive decline (Teychenne et al., 2012).

LEVEL OF INDIVIDUAL

At the level of the individual, it is important to actively work to educate yourself on understanding cognitive decline and its associated risk factor so that one can act on them. Education is crucial because, unlike some other illnesses, cognitive impairment is a result of the accumulation of pathological changes in the brain over several decades. As discussed earlier, effectively combating these risk factors involves a multi-prong approach which may overwhelm and discourage implementation. The first step in combating cognitive decline is being mindful of its symptoms so that disorders can be detected early. We recommend that individuals should keep track of any cognitive symptoms and complaints for themselves or family members. Physical and neurological examinations, along with laboratory test, should be performed routinely to identify potential risks that may not be obvious from observation. For long-lasting and sustainable changes to lifestyle, it is important to start small and strategically work to incorporate changes over time. Nutrition and physical activity are the two most important risk factors for cardiometabolic health and determinants of cognitive health in old age. We recommend adhering closely to the Mediterranean diet, which protects the brain from pathological age-associated changes and reduces the risk of cardiometabolic disorders that cause cognitive impairment. Similarly, to preserve cognitive function and reduce the risk of cardiometabolic disorders, routine exercise is also recommended. There is a lack of research on the type, duration, and frequency of exercise, which yields the greatest benefit; this is an area of future research. It is recommended for seniors to take part in cognitively stimulating leisure activities such as doing puzzles and playing board games or cards to maintain cognitive health. To alleviate stress, we recommend doing yoga and other relaxation exercises, ensuring adequate sleep, and maintaining a strong social network.

REFERENCES

CHAPTER 1

Agriculture.canada.ca. n.d. Overview of the food and beverage processing industry - agriculture.canada.ca. [online] Available at: <https://agriculture.canada.ca/en/canadas-agriculture-sectors/food-processing-industry/overview-food-and-beverage-processing-industry> [Accessed 23 August 2022].

Chen, X., Maguire, B., Brodaty, H., & O'Leary, F. (2019). Dietary Patterns and Cognitive Health in Older Adults: A Systematic Review. Journal Of Alzheimer's Disease, 67(2), 583-619. https://doi.org/10.3233/jad-180468

Fiolet, T., Srour, B., Sellem, L., Kesse-Guyot, E., Allès, B., & Méjean, C. et al. (2018). Consumption of ultra-processed foods and cancer risk: results from NutriNet-Santé prospective cohort. BMJ, k322. https://doi.org/10.1136/bmj.k322

Food and Agriculture Organization of the United Nations. (2019). Ultra-processed foods, diet quality, and health using the NOVA classification system. Rome. Retrieved from https://www.fao.org/3/ca5644en/ca5644en.pdf

Fox, M. (2012). Defining Processed Foods for the Consumer. Journal Of The Academy Of Nutrition And Dietetics, 112(2), 214-221. https://doi.org/10.1016/j.jand.2011.12.014

Jackson, P., Brembeck, H., Everts, J., Fuentes, M., Halkier, B., & Hertz, F. et al. (2018). Reframing convenience food. Springer International Publishing.

Luiten, C., Steenhuis, I., Eyles, H., Ni Mhurchu, C., & Waterlander, W. (2015). Ultra-processed foods have the worst nutrient profile, yet they are the most available packaged products in a sample of New Zealand supermarkets. Public Health Nutrition, 19(3), 530-538. https://doi.org/10.1017/s1368980015002177

Martin, C., Preedy, V., & Abbatecola, A. (2015). Diet and nutrition in dementia and cognitive decline. Academic Press.

Martínez Leo, E., & Segura Campos, M. (2020). Effect of ultra-processed diet on gut microbiota and thus its role in neurodegenerative diseases. Nutrition, 71, 110609. https://doi.org/10.1016/j.nut.2019.110609

Monteiro, C. (2009). Nutrition and health. The issue is not food, nor nutrients, so much as processing. Public Health Nutrition, 12(5), 729-731. https://doi.org/10.1017/s1368980009005291

Monteiro, C., Cannon, G., Levy, R., Moubarac, J., Louzada, M., & Rauber, F. et al. (2019). Ultra-processed foods: what they are and how to identify them. Public Health Nutrition, 22(5), 936-941. https://doi.org/10.1017/s1368980018003762

Parrish, A., 2014. What is a processed food?. [online] MSU Extension. Available at: <https://www.canr.msu.edu/news/what_is_a_processed_food> [Accessed 23 August 2022].

Sadler, C., Grassby, T., Hart, K., Raats, M., Sokolović, M. and Timotijevic, L., 2021. Processed food classification: Conceptualisation and challenges. Trends in Food Science & Technology, 112, pp.149-162.

The Nutrition Source. n.d. Processed Foods and Health. [online] Available at: <https://www.hsph.harvard.edu/nutritionsource/processed-foods/> [Accessed 23 August 2022].

van de Rest, O., Berendsen, A., Haveman-Nies, A., & de Groot, L. (2015). Dietary Patterns, Cognitive Decline, and Dementia: A Systematic Review. Advances In Nutrition, 6(2), 154-168. https://doi.org/10.3945/an.114.007617

Wahl, D., Cogger, V., Solon-Biet, S., Waern, R., Gokarn, R., & Pulpitel, T. et al. (2016). Nutritional strategies to optimise cognitive function in the aging brain. Ageing Research Reviews, 31, 80-92. https://doi.org/10.1016/j.arr.2016.06.006

Weinstein, G., Vered, S., Ivancovsky-Wajcman, D., Ravona-Springer, R., Heymann, A., & Zelber-Sagi, S. et al. (2022). Consumption of Ultra-Processed Food and Cognitive Decline among Older Adults With Type-2 Diabetes. The Journals Of Gerontology: Series A. https://doi.org/10.1093/gerona/glac070

CHAPTER 2

Baker, P., Machado, P., Santos, T., Sievert, K., Backholer, K., Hadjik-akou, M., Russell, C., Huse, O., Bell, C., Scrinis, G., Worsley, A., Friel, S., & Lawrence, M. (2020). Ultra-processed foods and the nutrition transition: Global, regional and national trends, food systems transformations and political economy drivers. Obesity Reviews, 21(12), e13126. https://doi.org/10.1111/obr.13126

Canada, H. (2002, September 27). History of Canada's Food Guides from 1942 to 2007 [Education and awareness;guidance]. https://www.canada.ca/en/health-canada/services/canada-food-guide/about/history-food-guide.html

Carpenter, K. J. (2003). A Short History of Nutritional Science: Part 3 (1912–1944). The Journal of Nutrition, 133(10), 3023–3032. https://doi.org/10.1093/jn/133.10.3023

Clay, R. (2017). The link between food and mental health. American Psychological Association, 48(8), 26.

Davis, C., & Saltos, E. (n.d.). Dietary Recommendations and How They Have Changed Over Time. 18.

Edwards, D. J. A. (1979). An Analysis of the Relationship between Nutrition and Psychological Health. South African Journal of Psychology, 9(3–4), 131–137. https://doi.org/10.1177/008124637900900310

Huebbe, P., & Rimbach, G. (2020). Historical Reflection of Food Processing and the Role of Legumes as Part of a Healthy Balanced Diet. Foods, 9(8), 1056. https://doi.org/10.3390/foods9081056

Jacka, F. N., Cherbuin, N., Anstey, K. J., Sachdev, P., & Butterworth,

P. (2015). Western diet is associated with a smaller hippocampus: A longitudinal investigation. BMC Medicine, 13(1), 215. https://doi.org/10.1186/s12916-015-0461-x

Koios, D., Machado, P., & Lacy-Nichols, J. (2022). Representations of Ultra-Processed Foods: A Global Analysis of How Dietary Guidelines Refer to Levels of Food Processing. International Journal of Health Policy and Management, 0. https://doi.org/10.34172/ijhpm.2022.6443

Lee, J. H., Duster, M., Roberts, T., & Devinsky, O. (2022). United States Dietary Trends Since 1800: Lack of Association Between Saturated Fatty Acid Consumption and Non-communicable Diseases. Frontiers in Nutrition, 8, 748847. https://doi.org/10.3389/fnut.2021.748847

Li, Y., Lv, M.-R., Wei, Y.-J., Sun, L., Zhang, J.-X., Zhang, H.-G., & Li, B. (2017). Dietary patterns and depression risk: A meta-analysis. Psychiatry Research, 253, 373–382. https://doi.org/10.1016/j.psychres.2017.04.020

Mozaffarian, D., Rosenberg, I., & Uauy, R. (2018). History of modern nutrition science—Implications for current research, dietary guidelines, and food policy. BMJ, 361, k2392. https://doi.org/10.1136/bmj.k2392

Polsky, J. Y., Moubarac, J.-C., & Garriguet, D. (2020). Consumption of ultra-processed foods in Canada. Health Reports, 31(11), 3–15. https://doi.org/10.25318/82-003-x202001100001-eng

Popkin, B. M., Adair, L. S., & Ng, S. W. (2012). NOW AND THEN: The Global Nutrition Transition: The Pandemic of Obesity in Developing Countries. Nutrition Reviews, 70(1), 3–21. https://doi.org/10.1111/j.1753-4887.2011.00456.x

Rangel, G. (2015, August 9). From Corgis to Corn: A Brief Look at the Long History of GMO Technology. Science in the News. https://sitn.hms.harvard.edu/flash/2015/from-corgis-to-corn-a-brief-look-at-the-long-history-of-gmo-technology/

Ríos-Hernández, A., Alda, J. A., Farran-Codina, A., Ferreira-García, E., & Izquierdo-Pulido, M. (2017). The Mediterranean Diet and ADHD in Children and Adolescents. Pediatrics, 139(2), e20162027. https://doi.org/10.1542/peds.2016-2027

Sukhatme, P. V. (2009). Size and Nature of the Protein Gap. Nutrition Reviews, 28(9), 223–226. https://doi.org/10.1111/j.1753-4887.1970.tb06236.x

Symbols, I. of M. (US) C. on E. of F.-P. N. R. S. and, Wartella, E. A., Lichtenstein, A. H., & Boon, C. S. (2010). History of Nutrition Labeling. In Front-of-Package Nutrition Rating Systems and Symbols: Phase I Report. National Academies Press (US). https://www.ncbi.nlm.nih.gov/books/NBK209859/

Weaver, C. M., Dwyer, J., Fulgoni, V. L., III, King, J. C., Leveille, G. A., MacDonald, R. S., Ordovas, J., & Schnakenberg, D. (2014). Processed foods: Contributions to nutrition. The American Journal of Clinical Nutrition, 99(6), 1525–1542. https://doi.org/10.3945/ajcn.114.089284

Weinstein, G., Vered, S., Ivancovsky-Wajcman, D., Zelber-Sagi, S., Ravona-Springer, R., Heymann, A., & Beeri, M. S. (2021). Consumption of ultra-processed food and cognitive decline among older adults with type-2 diabetes. Alzheimer's & Dementia, 17(S10), e055110. https://doi.org/10.1002/alz.055110

Welch, R., & Mitchell, P. (2000). Food processing: A century of change. British Medical Bulletin, 56, 1–17. https://doi.org/10.1258/0007142001902923

CHAPTER 3

Capetta, A. (2021, July 27). One major side effect of eating ultra-processed foods, says New Study. Eat This Not That. Retrieved August 22, 2022, from https://www.eatthis.com/news-side-effect-eating-ultra-processed-foods/

Edwin Kwong Research Fellow, Joanna Williams PhD Candidate, Phillip Baker Phillip Baker is a Friend of The Conversation. Research Fellow, Rob Moodie Professor of Public Health, &

Thiago M Santos PhD candidate. (2022, July 21). How big companies are targeting middle income countries to boost ultra-processed food sales. The Conversation. Retrieved August 22, 2022, from https://theconversation.com/how-big-companies-are-targeting-middle-income-countries-to-boost-ultra-processed-food-sales-166927

Ekstrand, B., Scheers, N., Rasmussen, M. K., Young, J. F., Ross, A. B., & Landberg, R. (2020, September 29). Brain Foods - the role of Diet in brain performance and health. OUP Academic. Retrieved August 22, 2022, from https://academic.oup.com/nutritionreviews/article/79/6/693/5912697

Fox, N. (2022, February 1). The many health risks of Processed Foods. LHSFNA. Retrieved August 22, 2022, from https://www.lhsfna.org/the-many-health-risks-of-processed-foods/#:~:text=Quicker%20to%20digest.,foods%20compared%20to%20unprocessed%20foods

Ifpri.org. (n.d.). Retrieved August 22, 2022, from https://www.ifpri.org/publication/changing-diets-urbanization-and-nutrition-transition

Klimova, B., Dziuba, S., & Cierniak-Emerych, A. (1AD, January 1). The effect of healthy diet on cognitive performance among healthy seniors – A mini review. Frontiers. Retrieved August 22, 2022, from https://www.frontiersin.org/articles/10.3389/fnhum.2020.00325/full

LaMotte, S. (2022, August 1). Cognitive decline linked to ultraprocessed food, study finds. CNN. Retrieved August 22, 2022, from https://www.cnn.com/2022/08/01/health/ultraprocessed-food-dementia-study-wellness/index.html

Link, K. (2021, May 3). How ultra-processed foods get US hooked - and how to resist. FoodPrint. Retrieved August 22, 2022, from https://foodprint.org/blog/ultra-processed-foods/#:~:text=Processed%20foods%20ignite%20our%20desire,they%20deliver%20more%20food%20quickly

MediLexicon International. (n.d.). Can ultra-processed foods affect cognitive performance? Medical News Today. Retrieved August 22, 2022, from https://www.medicalnewstoday.com/articles/can-ultra-processed-foods-affect-cognitive-performance

MediLexicon International. (n.d.). Processed Foods: Health risks and what to avoid. Medical News Today. Retrieved August 22, 2022, from https://www.medicalnewstoday.com/articles/318630#nutrients

NHS. (n.d.). Eating Processed Food . NHS choices. Retrieved August 22, 2022, from https://www.nhs.uk/live-well/eat-well/how-to-eat-a-balanced-diet/what-are-processed-foods/

Prado, E. L., & Dewey, K. G. (2014, April 1). Nutrition and brain development in early life. OUP Academic. Retrieved August 22, 2022, from https://academic.oup.com/nutritionreviews/article/72/4/267/1859597

Rapaport, L., Rapaport, L., Wahowiak, L., Groth, L., Thurrott, S., Upham, B., Ansel, K., Bedosky, L., & Blanton, K. (2022, August 5). Highly processed foods linked to accelerated cognitive decline. EverydayHealth.com. Retrieved August 22, 2022, from https://www.everydayhealth.com/diet-nutrition/highly-processed-foods-linked-to-accelerated-cognitive-decline/#:~:text=Too%20many%20daily%20calories%20from,than%208%2C000%20adults%20in%20Brazil

Tremblay, S. (2018, December 6). Does nutrition affect cognitive function? Healthy Eating | SF Gate. Retrieved August 22, 2022, from https://healthyeating.sfgate.com/nutrition-affect-cognitive-function-6132.html

U.S. Department of Health and Human Services. (n.d.). Your digestive system & how it works. National Institute of Diabetes and Digestive and Kidney Diseases. Retrieved August 22, 2022, from https://www.niddk.nih.gov/health-information/digestive-diseases/digestive-system-how-it-works

Young, L. (2019, May 30). We know highly processed food is bad for us. why do we keep eating it? - national. Global

News. Retrieved August 22, 2022, from https://globalnews.ca/news/5331125/ultra-processed-food-nutrition/

CHAPTER 4

Aday, S., & Aday, M. S. (2020, August 24). Impact of covid-19 on the Food Supply Chain. OUP Academic. Retrieved August 22, 2022, from https://academic.oup.com/fqs/article/4/4/167/5896496

Capozzi, F., Magkos, F., Fava, F., Milani, G. P., Agostoni, C., Astrup, A., & Saguy, I. S. (2021). A Multidisciplinary Perspective of Ultra-Processed Foods and Associated Food Processing Technologies: A View of the Sustainable Road Ahead. Retrieved August 22, 2022, from https://doi.org/10.3390/nu13113948

Chaarani, J. (2022, July 5). Fertilizer shortages could pinch 2023 food supply, says Ontario Federation of Agriculture | CBC News. CBCnews. Retrieved August 22, 2022, from https://www.cbc.ca/news/canada/london/russia-fertilizer-ontario-farm-1.6509279

Food Engineering. (n.d.). 2019 top 100 Food & Beverage Companies. Food Engineering RSS. Retrieved August 22, 2022, from https://www.foodengineeringmag.com/2019-top-100-food-beverage-companies

Government of Canada. (2015, December 14). Emerging Food Innovation: Trends and Opportunities. Government of Canada. Retrieved August 22, 2022, from https://agriculture.canada.ca/en/canadas-agriculture-sectors/food-processing-industry/trends-and-market-analysis-food-processing-industry/emerging-food-innovation-trends-and-opportunities

Government of Canada. (2021, October 28). Overview of the food and beverage processing industry. Government of Canada. Retrieved August 22, 2022, from https://agriculture.canada.ca/en/canadas-agriculture-sectors/food-processing-industry/overview-food-and-beverage-processing-industry

Hemanth. (2022, July 26). Verdict media limited. Food Processing Technology. Retrieved August 22, 2022, from https://www.foodprocessing-technology.com/features/top-ten-food-companies-in-2020/

Larson, S. (2022, July 1). The true costs of Processed Foods: Your Health, your planet. Escoffier. Retrieved August 22, 2022, from https://www.escoffier.edu/blog/sustainability/true-costs-processed-foods-health-planet/#:~:text=Factories%20can%20produce%20processed%20foods,consumers%20than%20fresh%20whole%20foods.

Processed food market - forecasts from 2020 to 2025. Research and Markets. (n.d.). Retrieved August 22, 2022, from https://www.researchandmarkets.com/reports/5067509/processed-food-market-forecasts-from-2020-to

CHAPTER 5

Anjum, I., Jaffery, S. S., Fayyaz, M., Wajid, A., & Ans, A. H. (2018). Sugar Beverages and Dietary Sodas Impact on Brain Health: A Mini Literature Review. Cureus, 10(6), e2756. https://doi.org/10.7759/cureus.2756

Gómez E. J. (2019). Coca-Cola's political and policy influence in Mexico: understanding the role of institutions, interests and divided society. Health policy and planning, 34(7), 520–528. https://doi.org/10.1093/heapol/czz063

Gómez E. J. (2021). The politics of ultra-processed foods and beverages regulatory policy in upper-middle-income countries: Industry and civil society in Mexico and Brazil. Global public health, 1–19. Advance online publication. https://doi.org/10.1080/17441692.2021.1980600

Mariath, A. B., & Martins, A. (2020). Ultra-processed products industry operating as an interest group. Revista de saude publica, 54, 107. https://doi.org/10.11606/s1518-8787.2020054002127

Moodie, R., Bennett, E., Kwong, E., Santos, T. M., Pratiwi, L., Williams, J., & Baker, P. (2021). Ultra-Processed Profits: The Political Economy of Countering the Global Spread of Ultra-Processed Foods - A Synthesis Review on the Market and Political Practices of Transnational Food Corporations and Strategic Public Health Responses. International journal of health policy and management, 10(12), 968–982. https://doi.org/10.34172/ijhpm.2021.45

CHAPTER 6

Gibney, M. J., Forde, C. G., Mullally, D., & Gibney, E. R. (2017). Ultra-processed foods in human health: a critical appraisal. The American journal of clinical nutrition, 106(3), 717-724.

Goodman, D. M. (2016). The McDonaldization of psychotherapy: Processed foods, processed therapies, and economic class. Theory & Psychology, 26(1), 77-95.

Harriden, B., D'Cunha, N. M., Kellett, J., Isbel, S., Panagiotakos, D. B., & Naumovski, N. (2022). Are dietary patterns becoming more processed? The effects of different dietary patterns on cognition: A review. Nutrition and Health, 02601060221094129.

Leddy, A. M., Weiser, S. D., Palar, K., & Seligman, H. (2020). A conceptual model for understanding the rapid COVID-19–related increase in food insecurity and its impact on health and healthcare. The American journal of clinical nutrition, 112(5), 1162-1169.

Leo, E. E. M., & Campos, M. R. S. (2020). Effect of ultra-processed diet on gut microbiota and thus its role in neurodegenerative diseases. Nutrition, 71, 110609.

Lustig, R. H. (2017). Processed food—an experiment that failed. JAMA pediatrics, 171(3), 212-214.

Relvas, G. R. B., Buccini, G. D. S., & Venancio, S. I. (2019). Ultra-processed food consumption among infants in primary health care in a city of the metropolitan region of Sao Paulo, Brazil. Jornal de pediatria, 95, 584-592.

Srour, B., & Touvier, M. (2020). Processed and ultra-processed foods: coming to a health problem?. International Journal of Food Sciences and Nutrition, 71(6), 653-655.

Taillie, L. S., Hall, M. G., Popkin, B. M., Ng, S. W., & Murukutla, N. (2020). Experimental studies of front-of-package nutrient warning labels on sugar-sweetened beverages and ultra-processed foods: a scoping review. Nutrients, 12(2), 569.

Weaver, C. M., Dwyer, J., Fulgoni III, V. L., King, J. C., Leveille, G. A., MacDonald, R. S., ... & Schnakenberg, D. (2014). Processed foods: contributions to nutrition. The American journal of clinical nutrition, 99(6), 1525-1542.

CHAPTER 7

Abushal, L. T., Salama, M., Essa, M. M., & Qoronfleh, M. W. (2021). Agricultural biotechnology: Revealing insights about ethical concerns. Journal of Biosciences, 46(3). https://doi.org/10.1007/s12038-021-00203-0

Klosse, P. R. (2019). The taste of a healthy and sustainable diet: What is the recipe for the future? Research in Hospitality Management, 9(1), 35–42. https://doi.org/10.1080/22243534.2019.1653590

Mezzacca, T. A., Anekwe, A. V., Farley, S. M., Kessler, K. A., Rosa, M. Q., Bragg, M. A., & Rummo, P. E. (2020). Ubiquity of Sugary Drinks and ProcessedFood Throughout Food and Non-Food Retail Settings in NYC. Journal of Community Health, 45(5), 973–978. https://doi.org/10.1007/s10900-020-00815-x

Monteiro, C. A., Cannon, G., Moubarac, J.-C., Martins, A. P. B., Martins, C. A., Garzillo, J., Canella, D. S., Baraldi, L. G., Barciotte, M., Louzada, M. L. da C., Levy, R. B., Claro, R. M., & Jaime, P. C. (2015). Dietary guidelines to nourish humanity and the planet in the twenty-first century. A blueprint from Brazil. Pub-

lic Health Nutrition, 18(13), 2311–2322. https://doi.org/10.1017/
S1368980015002165

Potthast, T., & Meisch, S. (2012). Climate change and sustainable
development Ethical perspectives on land use and food production (T. Potthast & S. Meisch, Eds.; 1st ed. 2012.). Wageningen
Academic Publishers. https://doi.org/10.3920/978-90-8686-
753-0

Rossi, J., Allen, J. E., Woods, T. A., & Davis, A. F. (2017). CSA
shareholder food lifestyle behaviors: a comparison across consumer groups. Agriculture and Human Values, 34(4), 855–869.
https://doi.org/10.1007/s10460-017-9779-7

Schermel, A., Mendoza, J., Henson, S., Dukeshire, S., Pasut, L.,
Emrich, T. E., Lou, W., Qi, Y., & L'abbé, M. R. (2014). Canadians' perceptions of food, diet, and health--a national survey.
PloS One, 9(1), e86000–e86000. https://doi.org/10.1371/journal.
pone.0086000

White., R (2020) on "For the Wild" podcast https://forthewild.
world/listen/rowen-m-white-on-seed-rematriation-and-fertile-
resistance-193

CHAPTER 8

Barrett, J. P., Olivari, B. S., Price, A. B., & Taylor, C. A. (2021).
Cognitive decline and dementia risk reduction: Promoting
healthy lifestyles and Blood Pressure Control. American Journal
of Preventive Medicine, 61(3). https://doi.org/10.1016/j.ame-
pre.2021.03.005

Baumgart, M., Snyder, H. M., Carrillo, M. C., Fazio, S., Kim, H., &
Johns, H. (2015). Summary of the evidence on modifiable risk
factors for cognitive decline and dementia: A population-based
perspective. Alzheimer's & Dementia, 11(6), 718–726. https://
doi.org/10.1016/j.jalz.2015.05.016

Chen, X., Huang, Y., & Cheng, H. G. (2012). Lower intake of vegetables and legumes associated with cognitive decline among

illiterate elderly Chinese: A 3-year cohort study. The Journal of Nutrition, Health & Aging, 16(6), 549–552. https://doi.org/10.1007/s12603-012-0023-2

Cooper, C., Sommerlad, A., Lyketsos, C. G., & Livingston, G. (2015). Modifiable predictors of dementia in mild cognitive impairment: A systematic review and meta-analysis. American Journal of Psychiatry, 172(4), 323–334. https://doi.org/10.1176/appi.ajp.2014.14070878

Dartigues, J. F., Foubert-Samier, A., Le Goff, M., Viltard, M., Amieva, H., Orgogozo, J. M., Barberger-Gateau, P., & Helmer, C. (2013). Playing board games, cognitive decline and dementia: A French population-based cohort study. BMJ Open, 3(8). https://doi.org/10.1136/bmjopen-2013-002998

Eyre, H. A., Acevedo, B., Yang, H., Siddarth, P., Van Dyk, K., Ercoli, L., Leaver, A. M., Cyr, N. S., Narr, K., Baune, B. T., Khalsa, D. S., & Lavretsky, H. (2016). Changes in neural connectivity and memory following a yoga intervention for older adults: A pilot study. Journal of Alzheimer's Disease, 52(2), 673–684. https://doi.org/10.3233/jad-150653

Gauvin, L., Richard, L., Kestens, Y., Shatenstein, B., Daniel, M., Moore, S. D., Mercille, G., & Payette, H. (2012). Living in a well-serviced urban area is associated with maintenance of frequent walking among seniors in the VOISINUAGE study. The Journals of Gerontology Series B: Psychological Sciences and Social Sciences, 67B(1), 76–88. https://doi.org/10.1093/geronb/gbr134

González-Gross, M., Marcos, A., & Pietrzik, K. (2001). Nutrition and cognitive impairment in the elderly. British Journal of Nutrition, 86(3), 313–321. https://doi.org/10.1079/bjn2001388

Hsu, T. M., & Kanoski, S. E. (2014). Blood-brain barrier disruption: Mechanistic links between western diet consumption and dementia. Frontiers in Aging Neuroscience, 6. https://doi.org/10.3389/fnagi.2014.00088

Lin, C.-H., Lin, E., & Lane, H.-Y. (2017). Genetic biomarkers on

age-related cognitive decline. Frontiers in Psychiatry, 8. https://
doi.org/10.3389/fpsyt.2017.00247

Nguyen, J. C., Killcross, A. S., & Jenkins, T. A. (2014). Obesity
and cognitive decline: Role of inflammation and vascular
changes. Frontiers in Neuroscience, 8. https://doi.org/10.3389/
fnins.2014.00375

Nooyens, A. C. J., Baan, C. A., Spijkerman, A. M. W., & Ver-
schuren, W. M. M. (2010). Type 2 diabetes and cognitive decline
in middle-aged men and women. Diabetes Care, 33(9), 1964–
1969. https://doi.org/10.2337/dc09-2038

Norton, S., Matthews, F. E., Barnes, D. E., Yaffe, K., & Brayne, C.
(2014). Potential for primary prevention of alzheimer's disease:
An analysis of population-based data. The Lancet Neurology,
13(8), 788–794. https://doi.org/10.1016/s1474-4422(14)70136-x

Pitchumoni, S. S., & Doraiswamy, P. M. (1998). Current status
of antioxidant therapy for alzheimer's disease. Journal of the
American Geriatrics Society, 46(12), 1566–1572. https://doi.
org/10.1111/j.1532-5415.1998.tb01544.x

Russ, T. C., Stamatakis, E., Hamer, M., Starr, J. M., Kivimäki, M.,
& Batty, G. D. (2013). Socioeconomic status as a risk factor for
dementia death: Individual participant meta-analysis of 86 508
men and women from the UK. British Journal of Psychiatry,
203(1), 10–17. https://doi.org/10.1192/bjp.bp.112.119479

Shakersain, B., Santoni, G., Larsson, S. C., Faxén-Irving, G.,
Fastbom, J., Fratiglioni, L., & Xu, W. (2015). Prudent diet
may attenuate the adverse effects of western diet on cognitive
decline. Alzheimer's & Dementia, 12(2), 100–109. https://doi.
org/10.1016/j.jalz.2015.08.002

Shatenstein, B., & Barberger-Gateau, P. (2015). Prevention of
age-related cognitive decline: Which strategies, when, and for-
whom? Journal of Alzheimer's Disease, 48(1), 35–53. https://doi.
org/10.3233/jad-150256

Taki, Y., Kinomura, S., Sato, K., Inoue, K., Goto, R., Okada, K.,

Uchida, S., Kawashima, R., & Fukuda, H. (2008). Relationship between body mass index and Gray Matter Volume in 1,428 healthy individuals. Obesity, 16(1), 119–124. https://doi.org/10.1038/oby.2007.4

Teychenne, M., Ball, K., & Salmon, J. (2012). Promoting physical activity and reducing sedentary behavior in disadvantaged neighborhoods: A qualitative study of what women want. PLoS ONE, 7(11). https://doi.org/10.1371/journal.pone.0049583

Wang, Y., Du, Y., Li, J., & Qiu, C. (2019). Lifespan intellectual factors, genetic susceptibility, and cognitive phenotypes in aging: Implications for interventions. Frontiers in Aging Neuroscience, 11. https://doi.org/10.3389/fnagi.2019.00129

Wang, Z., Pang, Y., Liu, J., Wang, J., Xie, Z., & Huang, T. (2020). Association of Healthy Lifestyle with cognitive function among Chinese older adults. European Journal of Clinical Nutrition, 75(2), 325–334. https://doi.org/10.1038/s41430-020-00785-2

Wen, X., Kong, K. L., Eiden, R. D., Sharma, N. N., & Xie, C. (2014). Sociodemographic differences and infant dietary patterns. Pediatrics, 134(5). https://doi.org/10.1542/peds.2014-1045